T0269470

Applied ethology

Proceedings of the 42nd Congress of the ISAE

Applied ethology
Addressing future challenges in animal agriculture

University College Dublin, Ireland

5-9 August 2008

Edited by:
Laura Boyle
Niamh O'Connell
Alison Hanlon

Wageningen Academic
P u b l i s h e r s

ISBN 978-90-8686-81-4

First published, 2008

Wageningen Academic Publishers
The Netherlands, 2008

Céad Míle Fáilte!

One hundred thousand welcomes to the 42nd Congress of the ISAE in Dublin!

The current food crisis is among many challenges facing agriculture in the world today. It is fitting then that the theme of this year's conference is 'Applied ethology: addressing future challenges in animal agriculture'. The basis for this theme is that knowledge and understanding of the principles and processes that govern farm animal behaviour can be transformed into practices that provide sustainable solutions to the challenges currently facing farm animal production systems.

The Scientific Programme reflects this theme but also demonstrates the diversity of research in Applied Ethology. Francoise Wemelsfelder will present the 'whole animal' approach in the Wood Gush Memorial Lecture which could help to make a more effective appraisal of the animal's perspective. While the five plenary papers range in topics from dominance in dogs to recent advances in the automatic collection of behavioural and physiological data. In addition the programme shows a strong interest in poultry welfare, which may be in anticipation of regulatory changes in the EU and thus reflect recent funding opportunities.

There are many similarities in terms of programme structure to previous ISAE Congresses, but we also tried to be innovative by introducing structured workshops. We hope that these will provide more opportunity for discussion and to exchange ideas.

The Social Programme and Excursions have also been an important focus for the Congress and we trust that they will provide great opportunities for delegates not only to see applied ethology in practice, but also to take time to unwind and enjoy Dublin and its surroundings.

As with any international event, this congress has been over three years in the planning. There are many colleagues that have helped us on our way. We would like to give a special acknowledgment to Francisco Galindo who took time before, during and after ISAE2007 to offer advice; Jack Murphy who helped in the early stages of preparation; Lorna Baird and numerous postgraduates who provided a list of tips on what to do and what not to do.

Alison Hanlon (University College Dublin, Ireland)
Laura Boyle (Teagasc, Ireland)
Niamh O'Connell (Agri-Food and Biosciences Institute, Northern Ireland)

Acknowledgements

Congress Organising Committee

Alison Hanlon (Chair)
Laura Boyle
Niamh O'Connell

Scientific Committee

Niamh O'Connell (Chair)
Laura Boyle (Chair)
Lorna Baird
Alan Fahey
Alison Hanlon
Jeremy Marchant-Forde
Sebastian McBride
Deborah Wells

Poster Sub-Committee

Alan Fahey
Keelin O'Driscoll

Workshop Sub-Committee

Anna Olsson
Laura Boyle
Niamh O'Connell

Assistant to Scientific and Congress Organising Committees

Keelin O'Driscoll

Professional Conference Organisers

Abbey Conference & Corporate
Patricia McColgan
Anne Griffin
Annalisa Scalavino

Design

Congress logo: Sebastian McBride
Congress website: Mark Boyd and Jack Murphy
Proceedings cover: Wageningen Academic Publishers

Referees

Andrew Fisher
Anna Valros
Anne-Marie de Passille
Anthony Podberscek
Bernadette Earley
Bob Elwood
Cassandra Tucker
Cathy Dwyer
Christine Nicol
Christoph Winckler
Dan Weary
Daniel Mills
Debbie Goodwin
Don Lay
Ed Pajor
Eva Baranyiova
Francoise Wemelsfelder
Hannah Buchanan-Smith
Hans Spoolder
Heleen Van de Weerd
Helen Edge
Helle Kristensen
Ian Sneddon
Jill Offer
John Mee

John Bradshaw
Jon Day
Jonathon Cooper
Joop Lensink
Kate Breuer
Kees van Reenen
Laura Hanninen
Marianne Bonde
Marie Haskell
Marina von Keyserlingk
Mark Rutter
Mike Appleby
Moira Harris
Nicola Marples
Nicola Rooney
Peter Hepper
Rick D'Eath
Sandra Edwards
Simon Turner
Stephanie Hayne
Stine Christiansen
Tine Rousing
Victoria Sandilands
Xavier Manteca

Congress Assistants

Sabine Conte
Gabriela Olmos
Jean Cosgrove
Emma Feerick
Paul Lightbody
Poppy Masterson
Catherine McCarney
Rayyan Mohamed
Cliona Mulvagh

Sponsors

science foundation ireland
fondúireacht eolaíochta éireann

AGRICULTURE AND FOOD DEVELOPMENT AUTHORITY

Department of
Agriculture, Fisheries and Food
An Roinn
Talmhaíochta, Iascaigh agus Bia

www.awionline.org

www.rspca.org.uk/sciencegroup

ciwf.org

World Society for the Protection of Animals

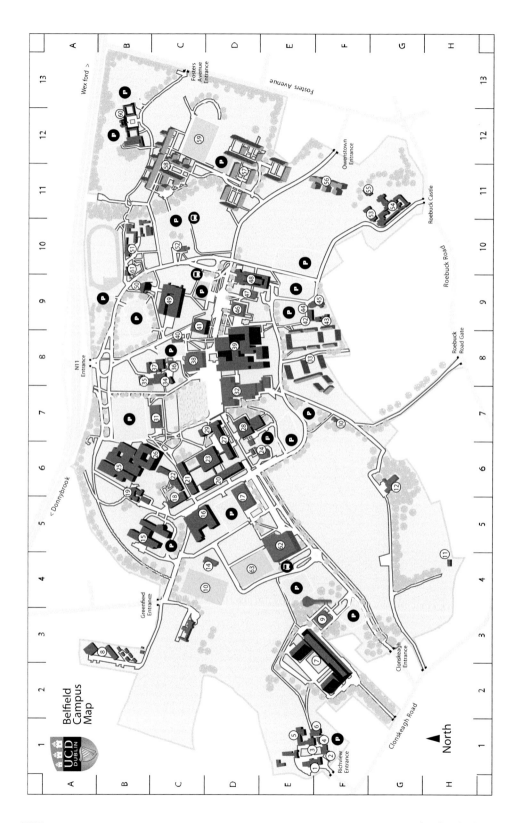

Belfield Campus Map

UCD DUBLIN

North

Applied ethology

Programme at a glance

Tuesday 5th		
14.00	Registration	
20.00	Welcome reception	

Wednesday 6th	Theatre 1	Theatre 2
08.00	Registration	
09.00	Congress Opening	
09.30	Wood-Gush Memorial Lecture	
10.30	Coffee/Poster session 1	
11.15	Plenary 1	
12.00	Behavioural indicators of health and welfare	Housing and environment I - Poultry
13.00	Lunch	
14.15	Housing and environment II - Cattle	Free communications I
15.15	Coffee/Poster session 2	
16.00	Workshops	
18.00	Finish	

Thursday 7th	Theatre 1	Theatre 2
08.45	Plenary 2	
09.30	Emotion, cognition and learning	Behavioural problems - Feather pecking
10.45	Coffee/Poster session 3	
11.30	Grazing behaviour	Free communications II
12.30	Finish	
12.45	Departure for excursions	

Friday 8th	Theatre 1	Theatre 2
09.00	Plenary 3	
09.45	Human - Animal interactions I	Social behaviour I
10.45	Coffee/Poster session 4	
11.30	Animal welfare	Breeding and genetics
12.45	Lunch	
14.15	Human - Animal interactions II	Mother and young
15.15	Coffee/Poster session 5	
16.00	ISAE AGM	
18.00	Finish	
19.30	Gala dinner	

Saturday 9th	Theatre 1	Theatre 2
09.30	Plenary 4	
10.15	Automated data collection methods	Stress
11.15	Coffee/Poster session 6	
12.00	Housing and environment III	Behavioural problems II
13.15	Lunch	
14.30	Plenary 5	
15.15	Social behaviour II	Feeding and behaviour - Pigs
16.15	Congress Closing	
16.30	Finish	
19.00	Farewell party	

O'Reilly Hall

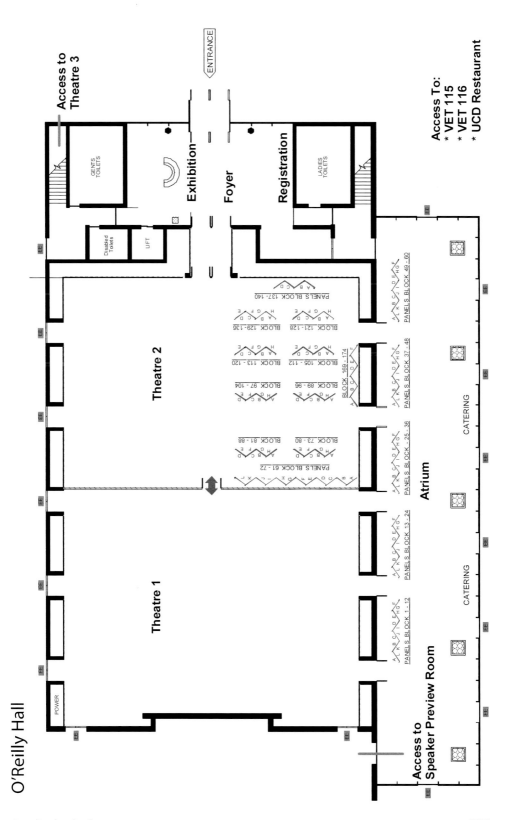

Applied ethology

Transport information

Transportation/Shuttle Service to/from Dublin Airport

Aircoach provides a 24-hour luxury coach service between Dublin Airport and Dublin city. Two routes operate, to Ballsbridge and Leopardstown, which serve Dublin's top hotels and places of business. This service is ideally suited to tourists, commuters and the general public looking for a high-quality, reliable, clean and efficient service.

Cost
Single ticket: €7.00 per person, payable on the coach.
Return ticket: €12.00 per person, payable on the coach.

Services
O'Connell Street, Grafton Street, Merrion Square, Pembroke Road, Ballsbridge, Simmonscourt Road, Leeson Street, St Stephens Green.

Visit www.aircoach.ie for more information on the AirCoach service.

Bus

Dublin Bus offers many routes throughout Dublin, including an AirLink express bus connecting Heuston and Connolly Rail Stations to Dublin Airport. Exact change is required for payment on the bus. Prepaid tickets can be purchased in most newsagents. The standard adult fare from Dublin City Centre to UCD is €1.70.

For more information go to www.dublinbus.ie

Taxi

Dublin airport is approx. 45 mins by taxi from the City Centre. Approximate price range for taxi trip = Euro 30-50.

David Wood-Gush Memorial Lecture

09:30 - 10:30 hours
Chairperson: Carol Petherick

Plenary 1

11:15 - 12:00 hours
Chairperson: Luiz Carlos Pinheiro Machado

Behavioural indicators of health and welfare

12:00 - 13:00 hours
Chairperson: Anna Johnson

Wednesday 6th August

Housing and environment I – Poultry

12:00 - 13:00 hours
Chairperson: Victoria Sandilands

Housing and environment II – Cattle

14:15 - 15:15 hours
Chairperson: Marie Haskell

Free communications I

14:15 - 15:15 hours
Chairperson: Harry Blokhuis

Workshop 1. Usefulness of resting behaviour as a tool for assessing the welfare of farm animals

16:00 hours
Chairperson: Sarah Chaplin

Wednesday 6th August

Workshop 2. Environmental enrichment: a valuable tool to improve the welfare of zoo animals?

16:00 hours
Chairpersons: Jessica Gimpel and Ivan Lozano

Workshop 3. Promoting applied ethology worldwide

16:00 hours
Chairperson: Marek Spinka

Workshop 4. The influence of genetics and breeding on farm animal welfare

16:00 hours
Chairperson: Bas Rodenburg

Workshop 5. Mother-offspring behaviour in polytocous animals

16:00 hours
Chairperson: Anna Olsson

Wednesday 6th August

Plenary 2

Theatre 1

Emotion, cognition and learning

Theatre 1

Behavioural problems I – Feather pecking

Theatre 2

Thursday 7th August

Grazing behaviour

11:30 - 12:30 hours
Chairperson: Mark Rutter

Free communications II

11:30 - 12:30 hours
Chairperson: Joop Lensink

Friday 8th August

Plenary 3

Human-animal interactions I

09:45 - 10:45 hours
Chairperson: Ed Pajor

Social behaviour I

09:45 - 10:45 hours
Chairperson: Hans Spoolder

Animal welfare

11:30 - 12:45 hours
Chairperson: Paul Koene

Friday 8th August

Breeding/genetics

11:30 - 12:45 hours
Chairperson: Per Jensen

Human-animal interactions II

14:15 - 15:15 hours
Chairperson: Simon Turner

Mother and young

14:15 - 15:15 hours
Chairperson: Stephanie Hayne

Theatre 2   **Page**

Saturday 8th August

Plenary 4

Automated data collection methods

Stress

Housing and environment III

12:00 - 13:15 hours
Chairperson: Anne Marie de Passille

Behavioural problems II

12:00 - 13:15 hours
Chairperson: Christoph Winckler

Saturday 8th August

Plenary 5

Theatre 1   Page

14:30 - 15:15 hours
Chairperson: Sean Kent

Social behaviour II

15:15 - 16:15 hours
Chairperson: Laura Hanninen

Theatre 1   Page

Feeding and behaviour – Pigs

15:15 - 16:15 hours
Chairperson: Rick D'Eath

Poster presentations

Each poster will be presented during 3 sessions. Authors are expected to stand by their poster during the sessions to which their poster is allocated.

Applied ethology   **XLVII**

L

David Wood-Gush Memorial Lecture

Curiosity, sentience, integrity: why recognising the 'whole animal' matters

F. Wemelsfelder, Scottish Agricultural College, Sustainable Livestock Systems, Sir Stephen Watson Building, Bush Estate, Penicuik, Edinburgh EH26 0PH, United Kingdom

Amongst David Wood-Gush's many contributions to ethological research were his studies of exploration and play in farm animals. Together with various colleagues he described how pigs show a pro-active, inquisitive curiosity in their surroundings, often accompanied by boisterous, playful bouts of scampering. Other scientists found that farm animals persist in activities such as food-searching and nest-building, even when these behaviours' functional end-points (food, nests) are already present. Such evidence illustrates that animals continuously and dynamically engage with their environment in a way not necessarily governed by concrete stimuli. In the scientific literature this capacity is addressed through concepts such as attention, responsiveness, information-processing, and decision-making, that all in some way denote an integrative level of behavioural organisation. Recent influential research models such as preference-testing and cognitive appraisal regard such integration as a crucial intervening variable, and accordingly provide more animal-centred assessments of welfare. However, given the prevailing mechanistic paradigm, animals in such research are conceived as complex systems to be analysed through sets of highly specified measures; the animal as such, i.e. the whole, integrated being, tends not to be directly addressed. In this presentation, I will explore why and how recognising the 'whole animal' matters to ethologists, particularly those interested in animal welfare. First of all, it is a matter of interpretative logic; we fundamentally rely on assessments of the 'whole animal' to attribute meaning and context to behavioural and physiological processes. It is the animal, not any (neuro)physiological system, who experiences pain, and for whom the things that happen create fear or pleasure. And so it is at this level we address the meaning of its expressions, to which specific measures of underlying mechanisms must ultimately refer to make sense. Thus we perceive animals as sentient in all the details of how they behave, and our research utterly depends on this perception. Furthermore, an awareness of animals as sentient beings affects us beyond the boundaries of scientific research. If, as a society, we did not see animals this way, our concern for their welfare would lose its moral urgency and remain an abstract thought. Animals have a presence, an integrity, both as individuals and collectively – this is why we care. These perceptions are the proverbial "elephant in the room" – they are so big and obvious that we tend to overlook them. However I believe that bringing the 'whole animal' more explicitly into ethological research will be highly beneficial; it will extend our understanding of animal sentience, and complement and strengthen other existing research approaches. I will illustrate this with a review of recent research on 'whole animal' assessment, also referred to as 'qualitative behaviour assessment' (QBA). This research is based on a methodology designed to assess an animal's expressive body language (e.g. as 'anxious' or 'content'), and has shown such assessments to be scientifically robust. For example, assessments of animal's body language were found to correlate well with, and enhance the interpretation of, other measures of behaviour and physiology. Such possibilities of mapping different assessment levels should make for better science, and for a more effective appraisal of the animal's perspective. Thus recognising the 'whole animal' has the potential to bridge disciplines, and to connect scientific assessment with wider ethical concerns.

Animal management and welfare, applied ethology and life as we know it

M.C. Appleby, World Society for the Protection of Animals, 89 Albert Embankment, London, SE1 7TP, United Kingdom

How animals are treated matters to animals, to people and to the environment, and is therefore extremely important to the future of the planet. There is increasing attention being given to animal management and welfare worldwide, and to the sciences that contribute to understanding of these topics, notably applied ethology. The benefits and potential benefits are relevant to many of the UN Millennium Development Goals, for both developed and developing countries, and are recognised by a number of intergovernmental organisations. They include benefits to: Human health – because good animal care reduces the risk of diseases transmissible to humans (zoonoses) and of food poisoning; the human-animal bond also has therapeutic effects; Social development – because people's attitudes and behaviour towards animals overlap with their attitudes and behaviour towards each other; Poverty and hunger reduction – because looking after animals properly improves their productivity and helps farmers to provide food for themselves, their families and their community; Disaster management – because animals are important for people's lives and livelihoods and must be considered in disaster preparedness and response; Environmental sustainability – because responsible animal management affects land use, climate change, pollution, water supplies, habitat conservation and biodiversity for the better. Applied ethology can and should make major contributions to these areas, which brings both opportunities and responsibilities for applied ethologists. Priorities need to be identified, and collaboration with other disciplines is essential – both with the disciplines specific to those priorities, and with social sciences such as economics to encourage the implementation of results.

Looking on the bright side of life: reward, positive emotions and animal welfare

L. Keeling[1], B. Algers[1], H. Blokhuis[1], A. Boissy[2], L. Lidfors[1], M. Mendl[3], R. Oppermann Moe[4], E. Paul[3], K. Uvnäs-Moberg[1] and A. Zanella[4], [1]Swedish University of Agricultural Sciences, Department of Animal Environment and Health, Box 7038/234, Uppsala/Skara, Sweden, [2]INRA, Herbivore Research Unit, Clermont-Ferrand -Theix Centre, F 63122 St Genès-Champanelle, France, [3]University of Bristol, Department of Clinical Veterinary Science, Langford House, BS40 5DU Langford, United Kingdom, [4]Norwegian School of Veterinary Science, Department of Production Animal Clinical Sciences, P.O. Box 8146, NO-0033 Oslo, Norway

In this paper we present a theoretical framework for the study of positive affective states in animals by integrating functional (emotional processes as proximate mechanisms that aid in achieving survival goals) and phenomenological (that emotions have valence and arousal dimensions) approaches. We propose that high arousal positive emotions (e.g. excitement, anticipation and desire) are associated with an appetitive motivational state. A subsequent consummatory motivational state is linked with sensory pleasure or 'liking' (e.g. pleasurable touch, hedonic taste) induced by an innate or acquired positive reinforcement from the resource. Then, a post-consummatory motivational state is associated with low arousal positive emotions such as 'satisfaction', 'relaxation' or 'relief' which function to aid recovery or restoration once the resource or goal is acquired. This concept of a 'Reward Cycle', wherein an organism passes through appetitive, consummatory and post-consummatory phases, links motivational and emotional theory. Our proposal, therefore, is based on a novel description in ethology of positive affective states as a multi-purpose reward cycle that can occur repeatedly across many different situations (e.g. feeding, drinking, sexual activity, play etc.). This approach is heuristically useful for generating hypotheses and research questions. In particular, it allows us to ask both mechanistic questions about the neuroanatomical and neurochemical pathways which determine the different phases of the reward cycle, and functional questions, particularly about the reward cycle's role in determining on-going behavioural and decision-making processes. For example, we hypothesise that a complete cycle is likely to result in the strongest positive affective state, and suggest investigating how manipulation of the cycle (e.g. by truncating / omitting certain phases or initiating a new cycle before the current one is completed) alters this affective state, and its behavioural and physiological indicators, and how repeated experience of complete/disrupted cycles maps on to the long-term assessment of positive state and mood.

Dominance in domestic dogs: useful construct or bad habit?

J.W.S. Bradshaw, E.J. Blackwell and R.A. Casey, University of Bristol, Centre for Behavioural Biology, DCVS, Langford BS40 5 DU, United Kingdom

Because the domestic dog Canis familiaris is descended from the wolf Canis lupus, it is widely assumed that its social relationships are derived from those of the wolf, and therefore that much of its social behaviour is driven by a motivation to enhance hierarchical positions. Thus aggression between dogs in a multi-dog household is often interpreted in terms of dominance rank, and as a consequence many clinical treatment protocols focus on "status reduction". However, there is increasing evidence that both the behavioural development and social cognitive abilities of domestic dogs have been substantially altered by domestication, calling into question the validity of the "wolf-pack" model. The agonistic and reproductive behaviour of feral dog packs does not resemble that of the wolf-pack; for example, although pair-bonding occurs, there is no evidence for suppression of reproduction within packs by "dominant" individuals of either sex. Agonistic interactions within stable groups of owned domestic dogs also do not follow wolf-type hierarchical patterns. For example, in one recent study of behavioural interactions within a permanent group of nineteen freely-associating neutered male dogs, although many confident/submissive interactions occurred, most were exchanged within a small number of dyads, and no overall hierarchy was evident. It may therefore be more parsimonious to interpret behavioural interactions between dogs as unique to each dyad, rather than there being any overall structure, hierarchical or otherwise, determining interactions across a whole group. This is consistent with the hypothesis that interaction between dogs is based largely on the specific experience of each individual about how others are likely to respond in different contexts. The so-called "dominant" dog (one that immediately selects aggression in a situation of social conflict) may simply be an individual that has previously learnt that aggression is an effective strategy for resolving social threats.

Recent advances in the automatic collection of animal behaviour and physiology

J.N. Marchant-Forde[1] and R.M. Marchant-Forde[2], [1]USDA-ARS, Livestock Behaviour Research Unit, 125 S. Russell St., West Lafayette, IN 47907, USA, [2]Purdue University, School of Veterinary Medicine, 625 Harrison St., West Lafayette, IN 47907, USA

The true assessment of an animal's state of being depends on the collection of refined, repeatable data, free from influence of the collection method or observer bias. The last few years have seen significant technical advances in automatic data collection pertaining to the animal and its environment. Although environmental data can be extremely useful, the most exciting advances have been with the collection of data from the animal itself. Early examples of automatic data collection include the radio-tracking of birds and feed recording in livestock. Since then, our capabilities have greatly increased so that we can now collect detailed information about aspects of an animal's behaviour, physiology, health and productivity. These data can be collected using a combination of non-invasive (environmentally mounted), more invasive (animal mounted) and most invasive (animal implanted) sampling systems. By using environmentally mounted systems, such as cameras, microphones and force plates, we can gather information on basic behaviour, such as activity and use of space. We might be able to detect lameness (force plate analysis), respiratory disease (acoustic analysis) and general sickness (activity and thermoregulatory behaviour analysis). Transponders mounted in ear tags or collars can supply information about feeding and drinking behaviour. We can measure frequency of feeder visits, amount of feed consumed, duration of feeding bout and we may record feed order (e.g. sows entering an ESF) or location, identity of neighbours and degree of social facilitation (e.g. cows feeding at a cattle bunk). Coupled with a weigh station or an automatic milking system, we might also collect growth data and milk output. Changes in the schedule of visits to the milking robot and automatic detection of milk quality can highlight cow health issues. Work with pedometers has advanced from the count-only type, which were mostly used to detect estrous behaviour in cattle, to more advanced accelerometers, that can log activity and posture-changing behaviour over time to highlight changes in circadian patterns and relation to reproductive state and health. Basic physiological information such as heart rate and respiration rate can be collected using external monitors. However, for more detailed physiology, we need to either implant telemetric devices or implant electrodes and catheters that enable us to then send information remotely or store the information on the animal for later retrieval. Combining collection systems will enable us to examine responses to specific stressors without human influence, greatly refining our data and expanding our knowledge.

Animal cognition, consciousness and animal welfare

M. Kiley-Worthington, Centre of Ecological Ethological Research & Education, La Combe Bezaudun sur Bine, 26460 Drome, France

Recent research on mammalian cognition indicates that many species mental attributes are considerable, and more akin to that of human primates than had previously been considered. There are few who doubt that other mammals are sentient, aware of others or being 'in the world'. In other words, they are to this degree conscious and have a subjectivity, an ontology. The existence of other species ontology opens up many important welfare questions. It has until recently either been denied or left aside by scientists, but there are now interdisciplinary approaches which are beginning to disentangle this intellectual stronghold. Of course, for other species as well as humans, the big problem of the first person experiences, or private lives remains. But there are ways proposed here on how to begin to outline another species ontology and consequently, consciousness. These include uniting and critically assessing both the third person approach, what has been called the objective stance, information from scientific experiments and observational studies, with the first person approach that involves a recognition and outlining for the subjective experiences of the world including emotions, shared experiences and folk knowledge. Such an approach has been called 'conditional anthropomorphism'. It poses many problems, but a summary of how it can throw rays of light on other species' ontology and consciousness is presented. The test case compares equines and elephants; a study that has been in train for 12 years to date. There are many applied aspects for such information. These include aid in making better informed welfare judgements of other species or individuals, and a better understanding of their particular quality of life.

The relationship between inactivity, reproductive performance and welfare state in mink

R. Meagher and G. Mason, University of Guelph, Animal & Poultry Science, 50 Stone Road East, Building #70, Guelph,Ontario N1G 2W1, Canada

In many species, individuals differ in their responses to being housed in barren environments: some become very stereotypic, while others become extremely inactive. Such inactivity could reflect either negative welfare states such as chronic fear or an analogue of human depression ('apathy'), or positive welfare states such as calmness. To determine whether inactivity potentially reflects poor welfare in mink, we examined the relationship between inactivity and reproductive performance in 350 individually-housed females (colour types: Black, Demi and Pastel) on a commercial farm. Behavioural phenotypes were determined through scanning observations conducted in the pre-feeding period for 4 days prior to the commencement of mating. Although most females were stereotypic, some were inactive on over 90% of the scans. Our reproductive measures included nest quality (reflecting an aspect of maternal care), scored by observers around the time of parturition; litter size at birth; and pre-weaning infant mortality. The relationships between these and inactivity were analysed using general linear models. High levels of inactivity in the nestbox predicted small litter sizes ($F_{1,330}=4.40$, $P=0.037$), and among multiparous females, it also predicted higher kit mortality between birth and weaning ($F_{1,202}=4.04$, $P=0.046$). Some signs of a link between inactivity and poor nest quality were also found, although they were not consistent over time. Stereotypic behaviour was significantly inversely correlated with inactivity, but was a less consistent predictor of reproductive performance. We are currently replicating this study on another 200 females; exploring the roles of depression-like states, fear, and/or excess body fat in our findings; and investigating whether different forms of inactivity (e.g. within the nest-box versus out in the open cage) vary in their correlates. In addition to the practical significance of identifying mink that are likely to breed poorly, extreme inactivity may have welfare implications if poor reproduction reflects depression-like states or fear.

On-farm welfare evaluation of group-housed sows

N.A. Geverink[1], B. Ampe[2], E. Struelens[1] and F.A.M. Tuyttens[1], [1]Institute for Agricultural and Fisheries Research (ILVO), Animal Sciences, Scheldeweg 68, 9090 Melle, Belgium, [2]Ghent University, Faculty of Veterinary Medicine, Department of Physiology and Biometry, Salisburylaan 133, 9820 Merelbeke, Belgium

To support the transition from individual to group housing of pregnant sows in Belgium, an observational study was started in 2005 to identify factors associated with welfare of group-housed sows. Forty pig farms with diverse group-housing systems were visited twice. Percentage of sows having saliva foam around their mouth as an indicator of sham chewing was scored. Skin damage of 20 sows per farm was scored on 9 parts of the body using a 0-4 scale and summed in a cumulative score. Parity number and days in gestation were recorded. Data on sham chewing were analysed using poisson regression with farm as a random effect and group size as offset. Data on skin damage were analysed with a mixed linear model with system, parity and gestation period as fixed effects, farm and group as random effects. Percentage of sows sham chewing was lower ($p<0.05$) in systems with ad libitum ($5.7\pm1.3\%$) or Vario-Mix® interval feeding ($6.8\pm2.1\%$) than in systems with free-access stalls ($33.29\pm3.8\%$) or trickle-feeding ($28.4\pm5.4\%$), and did not differ from systems with trough feeding ($42.3\pm13.3\%$). In electronic sow feeders (ESF, $13.9\pm1.8\%$) sham chewing was lower than in free-access stalls but higher than in ad-lib systems. Prevalence of sham chewing was 62% lower in systems with straw ($p<0.001$). In ESF-systems sows had more skin damage (9.9 ± 0.60) than in systems with free-access stalls (5.50 ± 0.44, $p<0.001$), ad libitum feeding (6.61 ± 0.74, $p<0.05$), trough feeding (5.79 ± 0.88, $p<0.01$), and – although not significant – Variomix systems (7.02 ± 0.81, $p=0.073$). Skin damage score decreased with parity number ($p<0.001$) and days in gestation ($p<0.001$). These results confirm the effectiveness of bulky feed or straw against sham chewing in group-housed sows in commercial units. Skin damage results indicate higher levels of aggression in ESF-systems, and show that gilts and young sows suffer relatively more from aggression in group housing.

Downcast donkeys and haggard horses: behavioural field indicators of poor welfare in working equids

C.C. Burn[1], T.L. Dennison[2], J.C. Pritchard[2] and H.R. Whay[1], [1]University of Bristol, Clinical Veterinary Sciences, Langford House, Bristol BS40 5DU, United Kingdom, [2]The Brooke, 21 Panton Street, London SW1Y 4DR, United Kingdom

Over 85% of the world's horses, donkeys and mules live in developing countries, used mainly for transporting goods and people by pack or by cart. Here we begin to evaluate the welfare relevance of behaviours recorded in a field assessment of working equids. Over a 5-year period, non-invasive behavioural and veterinary data were collected from 5481 donkeys, 4504 horses, and 858 mules across nine developing countries. Behaviours included the demeanours of undisturbed animals (alert or 'apathetic': ears lowered, little engagement with the environment), animals' responses to observer approach (none/ avoidance/ aggression/ friendly interest), and whether they accepted or rejected hand-to-chin contact. Relationships between the behaviours and body condition, lameness, and skin wounds were analysed. General linear models (Minitab, version 15) allowed factors including species, sex, age and location to be included; chi-squares were used for binary data. Thin animals (scoring ≤ 2 of 5) comprised 71% of the population, and were more apathetic ($F_{1, 10704}=199.3$; $P<0.001$), ignored the observer more ($F_{1, 10704}=40.7$; $P<0.001$), and rejected chin contact less ($F_{1, 10678}=5.26$; $P=0.022$) than fatter individuals. Lame animals avoided the observer less than sound ones ($\chi^2=163.9$; d.f.=1; $P<0.001$); 96% of equids were lame, but lameness showed no significant relationship with demeanour. Over 61% of equids had skin lesions, and those with more lesions were more apathetic ($F_{2, 10718}=3.98$; $P=0.019$) and yet more responsive to the observer ($F_{1, 10694}=12.6$; $P<0.001$) than those with few lesions. Lesions could perhaps have increased response to the observer through animals guarding against contact with the lesions. Results show that low body condition and lesions are associated with increased apathetic-like behaviour in a manner consistent with some working equines suffering poor welfare.

A comparison of the welfare of working elephants in India with elephants in zoos

M.J. Harris[1,2], C.M. Sherwin[1], S. Harris[3] and M. Pai[4,5], [1]University of Bristol, Centre for Behavioural Biology, Division of Clinical Veterinary Science, Bristol, BS40 5DU, United Kingdom, [2]Present address: Harper Adams University College, Animals Group, Newport, Shropshire, TF10 8NB, United Kingdom, [3]University of Bristol, School of Biological Sciences, Bristol, BS8 1UG, United Kingdom, [4]Wildlife Trust of India, A-220, New Friends colony, New Delhi, 110 025, India, [5]Present address: Clemson University, Department of Forestry and Natural Resources, 261 Lehotsky Hall, Clemson, SC, 29634, USA

During a study on the welfare of zoo elephants, we visited Kaziranga National Park in Assam to assess the welfare of 31 captive trained elephants kept under extensive conditions for tourist rides and forest duties. Data on the housing, husbandry, health and behaviour of these elephants were compared with those of all elephants in UK zoos and wildlife parks (N=41 Asian, N=36 African). We observed no stereotypies in working elephants, but their keepers reported that 29% showed stereotypic activity; we observed that 50% of Asian elephants in UK zoos (and 25% of Africans) exhibited stereotypies. Health problems were evident in both working and zoo elephants but were more prevalent in the working animals. Many working elephants had multiple signs of ill-health. For all body areas examined excluding feet, the percentage of working elephants with skin lesions was higher than that of zoo elephants. Overall foot health scores showed that foot problems were prevalent and did not differ in intensity between working and zoo elephants (2.32±2.04 and 1.98±1.63; mean ± s.d.; $F_{1,107}$=1.3, P=0.254). We devised an elephant gait scoring system to score 20 working elephants (aged 15 years and older) and 69 zoo elephants; 15% of working elephants had a normal gait and 10% were obviously lame, compared to 17% and 23% respectively for zoo elephants. A body condition score showed that of 25 working elephants and 69 zoo elephants, on a scale of 1 (very fat) to 5 (very thin), zoo elephants were significantly fatter than working elephants (2.03±0.44 and 3.25±0.53; $F_{1,93}$ =124.4, P<0.001). Welfare issues differed between the two populations: working Asian elephants had poorer skin health, but exhibited less stereotypies than zoo-housed Asians, and were less likely to be overweight than zoo-housed elephants.

A characterisation routine for improved animal welfare in pre-weaning mice

J.M. Marques[1], S.O. Ögren[2] and K. Dahlborn[3], [1]Instituto Gulbenkian de Ciência, Neurobiology of action, Rua da Quinta Grande 6, 2780-156 Oeiras, Portugal, [2]Karolinska Institute, Behavioural neuroscience, Retzius väg 8, Solna, 17177 Stockholm, Sweden, [3]Swedish University for Agricultural Sciences, Institute for Anatomy and Physiology, HVC, Ulls väg 27 D, F, 75007 Uppsala, Sweden

The increasing use of mutant mice raises ethical issues about producing/maintaining animal models which may undergo severe suffering. The need to monitor those animals for good development and welfare has been recognised. Mice' preweaning development is extremely fast. Monitoring during this stage allows detecting developmental abnormalities that cannot be observed during adulthood. We created a protocol for preweaning monitoring of mutant mice. This study aimed to evaluate its efficacy to detect development/welfare problems. 36 litters were screened, including four wild-type strains (C57BL/6, 129S6, BALB/C, B6CBAF1), two genetically modified lines (Nestin-PDGFB, GAD67-GFP) and one spontaneous mutant (Kv1.1-null mouse). Pups were monitored on postnatal days 1, 3, 7 and 14, according to a score-sheet comprising several morphological and reflexologic observations (e.g. vocalisations, righting-reflex, grip-strength). At weaning (day 21) a clinical examination was performed, evaluating physical (e.g. gait, posture, dehydration) and neurological condition (e.g. acoustic startle, visual placing, vertical pole, provoked biting) and emotional reactivity. Categorical data were analysed using Pearson Chi-square tests. For all other data, One-way ANOVA followed by Bonferroni post-hoc comparisons was used. For wild-type mice, development was strain dependent. During the first week of life, C57BL/6 mice had a lower weight gain (2.34 ± 0.06g) than all other strains (129S6: 2.93 ± 0.14g; BALB/C: 3.62 ± 0.09g; B6CBAF1: 2.84 ± 0.06g; $F(3,52)=33.83$, all $p<0.001$), and BALB/C had a higher weight gain than all other strains (all $p<0.001$). Less C57BL/6 pups (63%) developed the righting reflex on day 3, compared to other strains (129S6 and BALB/C: 90%, $\chi^2(1)=20.275$; B6CBAF1: 87%, $\chi^2(1)=15.36$; all $p<0.001$). Within litters of mutant mice, individual cases of abnormal development were found, including decreased body weight (n=24), absence of response to touch (n=21) and sound (n=7), abnormal gait (n=4), dehydration (n=8) and altered emotional reactivity (Nestin-PDGFB mice). The protocol is effective to detect welfare problems/development deviations that may require improved husbandry or humane endpoints.

Assessing walking style in commercially reared chickens using a force plate

V. Sandilands[1], S. Brocklehurst[2], L. Baker[1], D. Pearson[1], B. Thorp[3], R. McGovern[1] and N. Sparks[1], [1]SAC, West Mains Road, Edinburgh, EH9 3JG, United Kingdom, [2]BioSS, James Clerk Maxwell Building, Kings Buildings, Mayfield Road, Edinburgh, EH9 3JZ, United Kingdom, [3]Aviagen Ltd, Newbridge, Midlothian, EH28 8SZ, United Kingdom

What is perceived to be abnormal gait in chickens is an area for concern, because it may be painful and could restrict birds' access to resources. To objectively record walking style, we built a force plate (FP) 0.6 m long with load cells in each corner, which recorded foot step data onto a PC as birds walked across the plate. FP and conventional gait score (GS, birds assigned a score based on descriptions of walking style ranging from 0=walks with no impediment to 5=cannot sustain walking) data were assessed for their ability to predict post mortem (PM) results. GS results were reclassed as binary scores of GS<3 or GS>3, and a binary score for PM indicated presence/absence of any abnormalities or significant pathologies on either the left, right, or both, leg(s). Data were analysed using linear logistic models. Power calculations showed that a good predictor would requiresensitivity (Se) and specificity (Sp) >0.8. With FP data, we compared fits using a subset of 6 key measures and body weight because models that used all FP data were found to be over-fitting. 492 chickens from three flocks and two strains were assessed using FP and GS at 5 weeks of age. Of these, 191 birds of various GS (not representative of the flock GS overall) were culled for PMs. 51 birds (26.7%) had an abnormality (such as poor footpad scores) or significant pathology (such as tibial dyschondroplasia). Neither FP nor GS were good at predicting PM (FP 0.64 Se, 0.65 Sp, P<0.001; GS 0.58 Se, 0.63 Sp, P=0.010) but FP was marginally better. FP was better at predicting GS (Se 0.76, Sp 0.76, P<0.001). We concluded that FP and GS describe the way a bird walks, while PM describes leg health only.

The effects of morphine on thermal threshold and post-test behaviour in goldfish (*Carassius auratus*)

J. Nordgreen[1], J.P. Garner[2], A.M. Janczak[3], B. Ranheim[1], W.M. Muir[2] and T.E. Horsberg[1], [1]The Norwegian School of Veterinary Science, Dep Pharm Toxicol, Ullevålsvn 72, 0033 Oslo, Norway, [2]Purdue University, Dep Ani Sci, West Lafayette, 47907 IN, USA, [3]The Norwegian School of Veterinary Science, Dep Prod Anim Clin Sci, Ullevålsvn 72, 0033 Oslo, Norway

Standardised, repeatable, quantifiable measures of pain do not exist in fish. Therefore, we developed an apparatus to measure thermal threshold (TT) in fish. The aim was to test the hypothesis that fish respond to noxious heat with a baseline TT comparable to those of mammals, and to validate the model by testing the effects of morphine on TT and post-test behaviour. Sixteen fish were used. Half received 40 mg morphine/kg IM, half received 50 mg/kg in a cross-over study with saline control injections and wash-out 8 days. Home-tank behaviour (activity and hovering) was observed for 30 minutes at 24 hrs before the TT test, 30 min after the last TT test (+30) and 24 hours after the TT test. On the test day fish were anaesthetised and a small belt with a heater and sensor fitted. Heater temperature was increased at 0.9 °C/s, until the fish responded (the TT), and then immediately turned off. Two baseline TTs were measured. Then fish were injected with morphine/saline and TT tested 30 (TT30), 60 and 90 minutes after injection. Analyses used repeated measures ANOVA and post-hoc t-tests at α=0.05. Mean baseline TT was 37.9 °C (range: 32.5-42.9). The treatment-by-time interaction was significant ($F3.21=3.87$; $p=0.024$): 40mg/kg morphine increased TT30 significantly (to 41.4 °C) compared with baseline (39.6 °C) and saline (36.8 °C). Morphine at 50 mg/kg had no effects on TT. In the home tank, activity decreased at +30 following saline and 50 mg/kg morphine, but not 40 mg/kg morphine. To our knowledge, this study is the first to report TT in conscious fish. The TT is slightly lower than, but comparable to, those of mammals and can be affected by morphine. Implications for the debate over fish pain, and practical implications regarding analgesics for fish, will be discussed.

Effect of session length and reward size in an automated T-maze spatial discrimination task in mice

B.N. Gaskill, J.R. Lucas, E.A. Pajor and J.P. Garner, Purdue University, 125 South Russell, West Lafayette, 47907, USA

T-mazes are commonly used to test learning and memory in mice. However, they are time-intensive and sensitive to environmental variation. To address these issues we developed an automated maze attached to the home cage. Our first aim with this novel apparatus was to identify test parameters that maintain a mouse's accuracy of performance. We hypothesised that reward motivation would improve acquisition and maintain a constant performance. We predicted that larger rewards and shorter sessions would improve acquisition; and that smaller rewards and shorter sessions would maintain higher and less variable performance. Eighteen C57BL/6 mice (9 males and 9 females in groups of 3; 4-7 months old) performed a spatial discrimination task. The learning criterion was 8/10 consecutive trials correct. Mice were housed in a double-wide cage divided by a partition, which prevented non-testing cagemates from entering the maze while maintaining visual, olfactory and tactile contact with the test mouse. To manipulate reward motivation we used a 3x3 factorial design with 3 reward sizes (0.02, 0.04, or 0.08 g) and 3 session schedules (15, 30, or 45 minutes sessions with the same inter-session intervals). Each mouse had a total of 360 minutes of access to the maze per night, for two nights. Mouse was the experimental unit. Analysis used split-plot GLMs where linear contrasts tested for linear effects. No significant results were found in the number of trials or errors to criterion for either reward size or session length/interval. As predicted, reward size affected after-criterion average performance (GLM:$F_{2,11}$=12.6;p=0.001): performance improved as reward size decreased ($F_{1,11}$=25.1;p<0.001). As predicted, reward size affected after-criterion variability in performance (GLM:$F_{2,11}$=11.1;p=0.002): variability increased as reward size increased ($F_{1,11}$=17.9;p=0.001). Session length/interval did not affect any outcome. We conclude that smaller reward sizes improved maintenance and variability of performance, and neither these reward sizes nor session/interval lengths affected acquisition.

Visual and automated estrus detection methods in dairy cows at pasture and in cubicle accommodation

M. Palmer[1], G. Olmos[2], L. Boyle[2] and J.F. Mee[2], [1]University of Edinburgh, Royal (Dick) School of Veterinary Studies, Summerhall, Edinburgh, EH9 1QH, United Kingdom, [2]Moorepark Dairy Production Research Centre, Fermoy, Co Cork, Ireland

This study compared the efficiency (number of standing estrus detected divided by the number of ovulations) and accuracy (number of false standing estrus detected divided by the total number of standing estrus detected) of three methods of estrus detection in spring calving (mean calving date February 28th 2007) dairy cows at pasture (PASTURE, n=23) and in cubicle accommodation (HOUSED, n=23). For nine weeks, beginning 26th March, three estrus detection methods were simultaneously employed: visual observations of mounting and other sexual behaviours carried out for 20 minutes three times per day (VO), tail painting (TP) and radiotelemetry (HeatWatch). The dates of ovulation were determined from milk progesterone profiles measured three times per week. The breeding season began on 23rd April. In the PASTURE treatment there were 49 ovulations and detection efficiencies were: VO 59%, TP 65% and HeatWatch 69%. In the HOUSED treatment there were 46 ovulations and detection efficiencies were VO 20%, TP 26% and HeatWatch 37%. All methods were more efficient in PASTURE treatment (Chi-square tests, p<0.001). There was no difference between the efficiencies of the three methods within either treatment (p>0.10). Within each treatment there was no difference between the accuracy of the three methods (PASTURE: VO 97%, TP 94%, HeatWatch 97% and HOUSED: VO 100%, TP 92% and HeatWatch 77%). There was a tendency for HeatWatch to be more accurate in the PASTURE than HOUSED treatment (Fisher's exact test, p=0.076). During visual observation sessions, VO detected more mounts received by cows in standing estrus (median=2) than HeatWatch (median=0) (Sign test, N=12, M=5.5, p<0.01). Irrespective of the detection method, estrus detection was poor in cubicle-housed cows indicating that alternative approaches to estrus detection are needed for cows kept in such housing systems.

Welfare assessment through automated vocal type classification in piglets during different castration procedures and an open-field test

T. Schmidt[1], B. Bünger[2], T. Horn[1] and E. Von Borell[1], [1]Institute of Agricultural- and Nutritional Sciences, Martin-Luther-University Halle-Wittenberg, Dept. of Animal Husbandry and Animal Ecology, Adam-Kuckhoff-Str. 35, D-06108 Halle (Saale), Germany, [2]Friedrich-Loeffler-Institute, Federal Research Institute for Animal Health (FLI), Animal Welfare and Animal Husbandry, Dörnbergstrasse 25-27, 29223 Celle, Germany

The automatic detection of high frequency stress calls in pigs was proven to be a reliable tool in welfare research by previous investigations. In our approach, we focus on the analysis of the three different classes of piglets' vocalisations – grunting (G), squealing (Sq) and screaming (S). We hypothesise that this differentiation serves essential functions in pig communication and thus might reveal information about their well-being. In a first experiment, 3285 vocalisations from 42 piglets (German PIC; 7,13 and 19d) and in a second experiment 11089 vocalisations from 22 piglets (German Landrace; 24 to 27d) were analysed for 23 different vocal characteristics. A first discriminant function for the three vocal types (based on 13 sound characteristics) was derived from calibrated recordings made under laboratory-like conditions during castration procedures using the absolute sound energy content. A second discriminant function was derived from non-calibrated measurements of the relative energy content (based on 12 sound characteristics). The comparison of the two classifications showed an 86.7% agreement. The differences in the sound type distributions between the piglets castrated with (e.g. non-calibrated, S: 35.2%, Sq: 26.8%, G: 38.0%) and without anaesthesia (e.g. non-calibrated S: 46.3%, Sq: 16.7%, G: 37.0%) remained significant (Pairwise comparison, χ^2-test, P<0.05). The second classification allows for vocalisation analyses of animals under practical recording conditions. This was confirmed during an open-field test with piglets housed in two different farrowing systems. The proportion of screaming sounds in the open-field test was lower for piglets housed in a group-farrowing system (3.39±1.3%) than for those housed in a single-farrowing system (9.11±2.1%; χ^2-test, P<0.05). These results indicate that an increasingly stressful situation leads to an increase in the screaming proportion of vocalisations. We conclude that the classification of vocal types can add valuable information to studies using analyses of pig vocalisations for welfare assessment.

Effects of anaesthesia and analgesia on piglets' behaviour during four days following surgical castration

J.J. Zonderland and M. Kluivers-Poodt, Animal Sciences Group, Animal Production, Edelhertweg 15, 8219 PH Lelystad, Netherlands

Evidence exists that surgical castration induces pain not only during but also after the procedure. Local anaesthesia can suppress the pain during the castration procedure and analgesia is expected to suppress the pain afterwards. To assess the level of pain relief by using local anaesthesia and/or analgesia, the effect on piglet's behavioural response during four days after castration was measured. Within each of 24 litters, newborn male piglets housed in farrowing pens, were randomly allocated to one of six treatments: 1. Castration without anaesthesia or analgesia (CAST), 2. Castration after local anaesthesia with lidocaine (<st2: place w:st="on">LIDO</st2:place>), 3. Castration after local anaesthesia and administration of meloxicam (L+M), 4. Castration after administration of meloxicam (MELO), 5. Sham castration (SHAM) and 6. No handling or castration (NONE). For four days, each morning and afternoon, non-specific behaviour and specific pain-related behaviour were observed for 192 minutes using scan sampling at 12 minute intervals. Behavioural frequencies (percentage of total number of observations) were calculated. Residual estimated maximum likelihood (REML) was used for statistical analysis and student t-tests for possible treatment effects. Overall, LIDO piglets showed more tail wagging, which is considered to be a pain related behaviour, compared to other treatments (8.2% versus a range of 3.0 - 4.1%; $P<0.001$; including L+M piglets. During the first observation period after castration SHAM piglets were found more often without pain-related behaviour (74.0%) compared with LIDO (66.1%) and CAST (66.1%) piglets ($P<0.05$). Other behaviours like Walking, Manipulating and Awake Inactive sporadically showed an effect between treatments. This study showed a negative effect of lidocain on pain related behaviour in the days following castration. Meloxicam seemed to diminish these negative effects, although meloxicam alone had a limited positive effect on pain related behaviour after castration.

Investigating behavioural changes in young pigs infected with *Salmonella*

J. Higginson[1], J.T. Gray[2], T.M. Widowski[3], C.E. Dewey[1] and S.T. Millman[4], [1]University of Guelph, Department of Population Medicine, 50 Stone Road East, Guelph, ON N1G 2W1, Canada, [2]Des Moines University, Department of Microbiology & Immunology, 3200 Grand Avenue, Des Moines, IA 50312-4198, USA, [3]University of Guelph, Department of Animal & Poultry Science, 50 Stone Road East, Guelph, ON N1G 2W1, Canada, [4]Iowa State University, Veterinary Diagnostic & Production Animal Medicine/Biomedical Sciences, 2506 Vet Med, Ames, IA 50011-1250, USA

The objectives of this study were to determine if cleanliness, aggression, and exploratory behaviour would be altered by Salmonella infection in group-housed swine. This study was done in conjunction with a terminal microbiology project where twelve groups of five Landrace/ Yorkshire weaned pigs (n=60) were housed in separate biosecure rooms. One animal was randomly selected from each group as the seeder animal and given 10^7-10^8 CFU of *Salmonella typhimurium* orally on Day 0. Pigs were identifiable by individual markings and observers were blind to treatment. Cleanliness scores and fresh bite lesion scores were recorded daily from Day -1 to +6. Latency to approach a novel object, as a measure of exploratory behaviour, was performed using 4 different objects on Days -1, +1, +3 and +6, with order of presentation balanced between groups. Changes in cleanliness and lesion scores over time were analysed using logistic regression with a generalised linear mixed model in SAS, controlling for trial and pen. Latency to approach novel objects was compared using t-tests. Seeder pigs did not differ from penmates for cleanliness scores (0.70±0.10 vs. 0.67±0.06, p=0.83) or for bite lesion scores (0.21±0.06 vs. 0.14±0.03, p=0.15). All pigs approached novel objects significantly faster on subsequent days (p<0.05), however there was no significant difference between seeders and penmates. These results are in contrast to subtle behaviour changes by these seeder pigs which were detected using video analysis. In conclusion, newly weaned pigs that were inoculated with *Salmonella* did not display gross differences in behaviour relative to their healthy penmates.

Identifying changes in dairy cow behaviour to predict calving

H.M. Miedema[1], A.I. Macrae[1], C. Dwyer[2] and M.S. Cockram[3], [1]University of Edinburgh, Veterinary Clinical Sciences, Easter Bush Veterinary Centre, EH25 9RG, United Kingdom, [2]Scottish Agricultural College, Animal Behaviour and Welfare, Sir Stephen Watson Building, Bush Estate, EH26 0PH, United Kingdom, [3]University of Prince Edward Island, Sir James Dunn Animal Welfare Centre, 550 University Avenue, Charlottetown, C1A 4P3, Canada

Dystocia can result in significant economic loss due to cow and calf deaths, as well as reduced milk production and reproductive performance. Experienced stockmen use judgements based on physical and behavioural changes to recognise when cows may be about to calve, and offer assistance when required. With large herd sizes, and large numbers of cows per stockman, individual attention is difficult. This study aims to identify consistent behavioural changes which could potentially be used to predict calving. The behaviour of twenty multiparous Holstein-Friesian cows housed in straw-bedded pens, under continuous fluorescent or infra-red lighting, was recorded for 24 hours prior to the calf being expelled and for a 24-hour control period during late pregnancy. Continuous focal observations from video recordings were used to quantify the frequencies and durations of behaviours during 6-hour periods. For each 6-hour period, Wilcoxon tests were used to examine differences between behaviour during the calving and control periods. The frequencies of lying and tail raising were the most useful indicators of calving, as they showed consistent changes in the final 6-hour period during calving. During this period, lying frequency (number of lying bouts/6-hour period) was significantly higher ($p<0.001$) at calving (median=13, inter-quartile range=9-17) than during late pregnancy (median=4, IQR=3-5), and all cows showed an increase of ≥2 bouts. The frequency of tail raising also increased significantly ($p<0.001$) during the final 6 hours before calving (median=35, IQR=27-55) compared to the control period (median=5, IQR=3-7). Durations of lying, walking, eating and ground-licking, and number of walking bouts, did not show consistent changes at the time of calving. This study has shown that counting transitions between standing and lying, or tail raises, could potentially be useful for predicting calving.

Influence of farm factors on the occurrence of feather pecking in organic reared hens and their predictability for feather pecking in the laying period

M. Bestman[1] and P. Koene[2], [1]Louis Bolk Institute, Animal Husbandry, Hoofdstraat 24, 3972 LA Driebergen, Netherlands, [2]Wageningen University and Research Centre, Animal Sciences, Marijkeweg 40, 6709 PG Wageningen, Netherlands

Feather pecking is one of the most obvious welfare problems in laying hens, also in organic farming. Rearing circumstances play an important role in its development. We investigated the presence of feather damage (being the result of feather pecking) in organic rearing hens, the correlation between feather damage during the rearing period and during adult life and what farm factors were related to feather damage at young and adult age. We monitored 29 commercial flocks of rearing hens that were split up into 51 flocks of laying hens. Feather damage was scored during rearing at the ages of 7, 12 and 16 weeks and during laying at the age of 30 weeks. On the rearing as well as the laying farms, data were collected about the housing and management. We used logistic regression to analyse our data. Feather damage was seen in 13 out of 24 (54%) of the rearing flocks. The percentage of correct predictions concerning non feather damaged chicks not showing feather damage at adult age was 71.4%. When chicks showed feather damage, in 90.0% of the cases they did so during adult life. Logistic regression showed that a higher number of chicks being kept per square meter in the first 4 weeks of life, was associated with feather damage during the rearing period (Wald 5.434, df 1, sign 0.020, odds ratio 1.17). Moreover, the combination of not having substrate in the first 4 weeks and having much daylight in the age of 7 to 17 weeks, was a significant predictor of feather damage during the laying period (Chi square =13.957, df=3, p=0.003). These results can be used to improve animal welfare not only during rearing but also during later life. Although the observations were done on organic farms, the results can be applied for other non-cage systems too, where high density is also a common feature in chick rearing.

Effects of a mutation in PMEL17 on social and exploratory behaviour in chickens: possible relations with feather pecking

A.C. Karlsson[1], S. Kerje[2], L. Andersson[3], D. Nätt[1] and P. Jensen[1], [1]Linköping University, IFM Biology, Division of Zoology, Linköping University, SE-581 83 Linköping, Sweden, [2]University of Agricultural Sciences, Department of Animal Breeding and Genetics, Uppsala Biomedical Center, University of Agricultural Sciences, SE-751 24 Uppsala, Sweden, [3]Uppsala University, Department of Medical Biochemistry and Microbiology, Box 582, SE-751 23 Uppsala, Sweden

Colour changes in fur and plumage are important domestication effects, and the PMEL17 protein affects tissue pigmentation. PMEL17 mutations cause white plumage in chickens, while the wild type shows a pigmented plumage. Earlier studies have shown that a mutation in PMEL17 protects against feather pecking. This might be related to differences in social behaviour between PMEL17 genotypes. We studied behaviour and gene expression in brains of homozygous PMEL17 genotypes, using chickens originating from an advanced White Leghorn x red jungle fowl intercross. Social behaviour was studied in three different behavioural tests in two experimental replicates (replicate 1: 19 white and 6 wild-types; 2: 26 white and 24 wild-types). Data were analysed by independent-samples T-test (SPSS 12.0.1) and there were significant differences between the genotypes in both replicates, but the direction of the results was not consistent. For example, in replicate 1, wildtypes were vocalising more (57.1±4,4 vs. 20.0±4,8% of observations; $p<0,001$) and spending more time in the social zone (70.4±7.2 vs. 34.1±9.1% of observations; $p<0,01$) of a social versus foraging test whereas no or opposite tendencies were found in replicate 2. Hence PMEL17 effects on social behaviour may be less important than hypothesised with respect to feather pecking victimisation. In an exploration test, wild type chickens were more explorative (6 vs. 0 explorative zone entries; $p<0.05$), and this may be a more important risk factor for being exposed to feather pecking. Gene expression studies showed no PMEL17 expression in brain, so the genotype differences must depend on extra-neural gene expression. We conclude that PMEL17 may affect social behaviour and explorative behaviour, but more research is needed before this can be clearly related to the increased risk of being a feather pecking victim.

Responses to bitter and sweet tasting feathers in laying hens

A. Harlander-Matauschek, F. Wassermann and W. Bessei, Department of Small Animal Ethology and Poultry Science, Garbenstr. 17 - 470c, 70599 Stuttgart, Germany

Recent investigations have shown that feather pecking birds eat plucked feathers. Studies about taste preferences in laying hens showed that birds like the flavour of sweet tasting sucrose and dislike bitter tasting quinine sulphate. Consequently, we hypothesised that laying hens, which are exposed to bitter tasting feathers, will avoid later pecking, plucking and eating feathers from other birds and do the opposite when the birds are exposed to sweet tasting feathers. Forty eight Lohmann Brown hens with a high propensity to feather pecking were picked out of group housed laying hens. The birds were transferred to individual cages and fed a pelleted diet ad libitum. After one week of habituation to the experimental situation 12 birds were offered ten feathers soaked in a bitter tasting 4% quinine sulphate solution (Q) and the other 12 birds were offered feathers in 4% sucrose solution (S) on a daily basis over a period of ten days. The other 24 birds were kept as a control (C) without access to feathers. After a ten day feather feeding period, three groups of four S and four C birds each and three groups of four Q and four C birds each were created. Bouts of severe and gentle feather pecking behaviour were recorded over a period of eight days. The number of S feathers eaten increased throughout the feather feeding phase, whereas Q feathers were avoided ($P<0.001$). Birds which were offered Q feathers in the feather feeding phase, showed significantly less severe feather pecking than S ($P<0.007$) and C birds ($P<0.04$). The groups did not differ in their gentle feather pecking bouts. The results clearly showed that the bitter tasting quinine sulphate was the signal the birds used to avoid pecking at conspecifics.

Factors affecting the development of injurious pecking in flocks of free range and organic laying hens

S.L. Lambton[1], T.G. Knowles[1], C. Yorke[2] and C.J. Nicol[1], [1]University of Bristol, Farm Animal Science, Langford House, Langford, North Somerset, BS40 5DU, United Kingdom, [2]Stonegate Farmers, The Old Sidings, Lacock, Chippenham, Wiltshire, United Kingdom

Injurious pecking remains one of the biggest problems challenging free-range egg producers, with both economic implications for the farmer and welfare implications for the birds. This prospective epidemiological study investigated the development of injurious pecking on 61 free-range and organic UK farms (111 flocks). Flocks were visited at 25 and 40 weeks, when rates of gentle and severe feather pecking (GFP and SFP respectively), vent pecking, and levels of feather damage, were recorded. Environmental and management data were collected for each flock. Factors affecting the development of these behaviours were modelled using the multilevel modelling program, MLwiN. GFP was observed in 89.2% and 73% of flocks at the 1st and 2nd visits, with a mean rate of 0.0129 bouts/bird/min. GFP rates decreased with age (Z=4.68, p<0.001), increasing temperature inside the laying house (Z=2.13, p<0.001), increased range use (Z=2.24, p=0.025), and were higher in beak trimmed flocks (Z=5.82, p=0.007), flocks without perches (Z=1.99, p=0.047), and flocks with soils as litter (Z=2.26, p=0.024). SFP was observed in 68.5% and 85.6% of flocks at 1st and 2nd visits, with a mean rate of 0.0191 bouts/bird/min. SFP rates increased with age (Z=2.96, p=0.003), and decreased with range use (Z=3.23, p=0.001). Rates were also highest in non-beak trimmed flocks (Z=3.90, p<0.001), flocks observed to be feather pecking when they arrived on farm (Z=2.45, p=0.014), and flocks fed pelleted food (Z=2.81, p=0.005). Feather damage was lower in beak trimmed flocks (Z=2.41, p=0.016), and showed quadratic relationships with severe feather pecking (Z=2.97, p=0.003) and vent pecking (Z=5.66, p<0.001), both of which were positive over the observed ranges for the behaviours. We conclude that severe feather pecking, and therefore feather damage, may be reduced by feeding mashed food and increasing range use.

Injurious pecking in laying hens: risk factors during rearing and laying phase in conventional and organic flocks

M. Staack[1], B. Gruber[2], C. Keppler[1], K. Zaludik[2], K. Niebuhr[2] and U. Knierim[1], [1]University of Kassel, Farm Animal Behaviour and Husbandry, Nordbahnhofstr. 1 a, 37213 Witzenhausen, Germany, [2]University of Veterinary Medicine Vienna, Institute of Animal Husbandry and Welfare, Veterinärplatz 1, 1210 Vienna, Austria

As part of the welfare quality project an epidemiological approach was used to explore the risk factors for injurious pecking in laying hens kept in non-cage systems during both the rearing and laying period. During one day visits to 23 organic and 27 conventional rearing units and to two subsequent laying units of each rearing unit in Austria and Germany, data on potential risk factors were collected by recording housing details, interviewing farmers about management, and weighing random samples of 30 hens. Regression tree analysis was carried out with the dependent variable 'prevalences of injuries' which had been determined from the 30 hens. The average percentage of laying hens with injuries did not differ significantly between organic (mean 22.8%) and conventional (mean 13.2%) flocks (Mann-Whitney-U-test, p=0.544). Identified risk factors for prevalences of injuries explained 64% (organic) and 56% (conventional) of total variance between the 46 and 54 flocks, respectively. Risk factors (proportions of explained variance in brackets) for organic flocks included little elevated perch space/pullet (57%), relatively high hen weight (19%), ammonia content during rearing (17%) and little drinking space/hen (7%). In organic flocks 'age of hens in days at scoring' was a confounding factor. For conventional flocks the risk factors relatively low pullet weight (31%), high stocking density of pullets (30%), little feeding space/pullet (22%), no access to covered outside run during laying (8%), high ammonia content during rearing (7%), higher age of hens when moved to laying unit (1%) were identified. When calculating the contribution of risk factors during rearing relative to those during laying (as proportion of the total explained variation on the basis of the respective sums of squares), 91% (conventional) and 74% (organic) of variation was explained by rearing factors. The identified factors should be considered when prevention strategies against injurious pecking are developed.

Prediction of tail-biting from behaviour of pigs prior to outbreaks

P. Statham[1], L.E. Green[2], M. Bichard[1] and M. Mendl[1], [1]University of Bristol, Clinical Veterinary Science, Langford House, Langford, North Somerset, BS40 5DU, United Kingdom, [2]University of Warwick, Department of Biological Sciences, Coventry, CV4 7AL, United Kingdom

The ability to predict tail-biting outbreaks could be a useful tool enabling farmers to intervene and thus improve pig welfare. Over 700 pigs were studied from birth to slaughter, housed indoors with natural ventilation on solid floors. At weaning, twenty groups were formed from three litters each. Fourteen outbreaks of tail-biting occurred, classified as severe (blood in pen, N=6) or underlying (low levels of tail damage, N=8). Behaviour prior to each outbreak was compared to control groups where no tail-biting occurred. We hypothesised that pigs would show increased activity, tail-orientated behaviour, and 'tucked-under' tail postures prior to outbreaks. The proportion of pigs performing different behaviours in severe tail-biting (TB) and matched control (C) groups during the four days prior to an outbreak was recorded on video. Repeated measures GLMs showed that activity was higher in tail-biting groups, with more pigs standing (TB=0.21±0.02,C=0.17±0.01, $F_{1,5}$=7.744,p<0.05) and fewer pigs sitting (TB=0.01±0.001,C=0.03±0.005, $F_{1,5}$=12.322,p<0.05) or lying inactive (TB=0.51±0.02,C=0.59±0.02, $F_{1,5}$=7.549,p<0.05). Pigs were observed directly when seven and eleven weeks of age, enabling pre-outbreak behaviours of all severe (S), underlying (U) and no (N) outbreak groups to be compared. Repeated measure GLMs or non-parametric equivalents showed higher levels of tail interest (manipulating/nosing) in no outbreak groups than in severe outbreak groups (N=0.74±0.20,U=0.60±0.17,S=0.19±0.06, $F_{2,17}$=4.933,p<0.05), while the opposite was true for more vigorous tail-biting behaviour (N=0.00±0.00,U=0.00±0.00,S=0.00±0.16, χ^2=8.454,df=2,p<0.05). Fewer tail 'tucked-under' postures were observed in groups without outbreaks (N=0.00±0.00,U=1.40±4.25,S=0.90±1.85, χ^2=11.080,df=2,p<0.01). Our results suggest that increased activity and 'tucked-under' tail posture may occur prior to tail-biting outbreaks. Interestingly, damaging tail biting may increase prior to an outbreak, but the opposite seems to be the case for tail interest. Further work is needed to understand the relationship between different forms of tail contact. With further research and advances in technology, behavioural changes may be used to predict tail-biting outbreaks on commercial farms.

Cannibalism in clean cages: effects of cleaning frequency on the welfare of breeding rats

C.C. Burn[1] and G.J. Mason[2], [1]University of Bristol, Clinical Veterinary Sciences, Langford House, Bristol BS40 5DU, United Kingdom, [2]University of Guelph, Animal and Poultry Science, Guelph, Ontario N1G 2W1, Canada

Cleaning the cages of breeding rats regularly is important for their health, but it disrupts the nest and removes olfactory signals important for parental care, such as pheromones deposited in the bedding that help prevent co-habiting males killing pups and that keep pups in the nest. We investigated how different cage-cleaning frequencies affect breeding rat health and welfare. Cages were cleaned twice-weekly, once-weekly or every two weeks (18 cages/group), replicated in two buildings within a large commercial breeding facility. Reproductive output, pup mortality, pup sex ratios, parental chromodacryorrhoea (a stress-related harderian gland secretion) and ammonia levels were monitored for the entire 36-week reproductive period. Data were analysed using generalised linear models (R, version 2.4 and Minitab, version 14), which could include the two buildings and, where relevant, repeated measures. Frequent cage-cleaning had no clear benefits under these clean barrier conditions, showing no significant reductions in ammonia levels, or effects on health or overall pup mortality. However, cannibalistic behaviour was slightly but significantly more likely with more frequent cage-cleaning ($z=2.15$; $n=106$; $P=0.032$). This was because the risk of cannibalism was greater if cleaning occurred within the first two days of birth. Possible mechanisms behind these effects include that cleaning might induce premature births and hence weaker pups, or it could stress the parents through acute auditory, olfactory or physical disturbance at a time when pups are particularly vulnerable. Finally, the cage-cleaning frequency producing most pups differed between the two buildings ($F_{2, 100}=4.30$; $P=0.016$), an interactive effect supporting previous findings that same-strain rodents' phenotypes can differ with environment. To summarise, we suggest that cage-cleaning regimes for breeding rats should minimise noise, dissemination of unfamiliar conspecific odours, and physical disturbance during very late pregnancy and the first few days following birth.

Individuality and restricted feeding affect stereotypy in beef cattle under confinement

R.J. Kilgour[1], T. Ishiwata[2], G.J. Melville[1] and K. Uetake[2], [1]New South Wales Department of Primary Industries, Agricultural Research Centre, PMB 19, Trangie NSW 2823, Australia, [2]Azabu University, Department of Animal Science and Biotechnology, 1-17-71 Fuchinobe, Sagamihara 229-8051, Japan

One of the purported indicators of poor welfare in animals in close confinement is stereotypic behaviour. An example of stereotypy in beef cattle under confined conditions is tongue rolling. Restricted feeding has been shown to increase the incidence of tongue rolling and may reflect frustration through the thwarting of normal feeding behaviour. Individual animals differ in the incidence of performance of tongue rolling, so it may be possible to identify potential tongue rolling animals before the imposition of confinement. Here, 106 Wagyu-Holstein steers aged approximately one year underwent a five-minute fear of humans test. The animals were then housed in nine pens, each of which measured 6m x 9.5m. There were 11 or 12 animals per pen, and in six of the pens, the animals were fed twice daily, whereas in the other three they were fed ad libitum. The behaviour of the animals in the pens was measured using 15-minute instantaneous scans for a total of three days. A principal components analysis of the fear of humans test identified two components, which were named "attraction to the human" and "general agitation". "Attraction to the human" explained 70.0% of the variance and "General agitation" explained 16.1% of the variance. The incidence of tongue rolling was higher under the restricted feeding system (2.4% vs 0.4%, $\chi^2_{1df} = 106.3$, P<0.001). In the animals under the restricted feeding regimen there was also a significant effect of attraction to the human (P<0.05) on tongue rolling in that animals that avoided the human performed a higher rate of tongue rolling. These relationships were not significant under the ad libitum feeding regimen. These results indicate that tongue rolling may result from restricted feed intake and that animals that are more willing to approach the human in the fear of humans test cope better with this feeding system.

Stereotypy is heritable and is associated with increased reproductive output in captive striped mice (Rhabdomys)

M.A. Jones[1], M.C. Van Lierop[1], G.J. Mason[2] and N. Pillay[1], [1]University of the Witwatersrand, School of Animal, Plant, and Environmental Sciences, Private Bag 3, WITS, 2050, South Africa, [2]University of Guelph, Department of Animal and Poultry Science, 50 Stone Road East, Guelph, Ontario N1G 2M7, Canada

Stereotypic behaviour is common in captivity, but is not observed in wild or free-ranging animals. We investigated the relationship between stereotypy and reproductive output in captive striped mice, Rhabdomys, and tested for the genetic transmission of the behaviour. Mice (n=120) were assigned to one of four treatment groups (non-stereotypic parents; stereotypic mother, non-stereotypic father; non-stereotypic mother, stereotypic father; stereotypic parents), and various measures of reproductive output, as well as offspring stereotypy prevalence, were recorded. On most measures, reproductive output (e.g., total number of offspring; $F_{3,47}=8.82$; $p<0.001$; General Linear Model) of stereotypic females (but not stereotypic males) was greater than in non-stereotypic mice (7.04 ± 0.26 v. 4.47 ± 0.30). Whilst this could be taken as evidence for enhanced coping in stereotypic striped mouse mothers, other work has indicated that the improved reproductive success of these females is indirectly mediated via a nutritional mechanism (i.e., greater protein intake). In terms of inheritance, the incidence of stereotypy was five times higher in the offspring of stereotypic than non-stereotypic females ($51.89\pm1.92\%$), regardless of whether the male was stereotypic, and three times greater in offspring sired by stereotypic males paired with non-stereotypic females ($36.34\pm5.83\%$) than in offspring from non-stereotypic parents ($10.36\pm3.22\%$; $F_{3,46}=22.52$; $p<0.001$). This shows that stereotypy has a strong genetic basis, even though the significantly greater maternal than paternal influence on the development of stereotypy supports the additional role of epigenetic factors in the development of the behaviour. Overall, because (1) stereotypic females breed more successfully than non-stereotypic mice, and (2) genetic variance underlies the trait, unintended artificial selection processes are likely to shift the evolutionary trajectory of captive-bred Rhabdomys. If these two criteria are met in other species, stereotypy will be associated with the phenotypic and genotypic divergence of captive stock from wild populations, and its performance could have implications for captive breeding for conservation.

Gastric ulceration in mature horses with history of crib-biting

C.L. Wickens[1], C.A. McCall[2], S. Bursian[1], R.R. Hanson[3], S.J. Holcombe[4], J.S. Liesman[1], W.H. McElhenney[5], C. Heleski[1] and N.L. Trottier[1], [1]Michigan State University, Animal Science, Anthony Hall, 48824 East Lansing, USA, [2]Auburn University, Animal Sciences, Animal Sciences, 36849 Auburn, USA, [3]Auburn University, Clinical Sciences, Large Animal Teaching Hospital, 36849 Auburn, USA, [4]Michigan State University, Large Animal Clinical Sciences, College of Veterinary Medicine, 48824 East Lansing, USA, [5]Tuskegee University, Animal Sciences, Animal Sciences, 36088 Tuskegee, USA

Gastrointestinal irritation has been implicated in crib-biting (CB) in horses. We hypothesised that horses exhibiting CB have more gastric mucosal damage than non-cribbing (NCB) horses. Eighteen horses (9 CB and 9 NCB) were used to determine the prevalence and severity of gastric mucosal damage and the effect of feeding on circulating gastrin concentrations in CB and NCB horses. Horses were maintained on coastal bermudagrass pasture with free access to bermudagrass hay and water and twice daily delivery of a pelleted (10% protein) diet. Endoscopic examinations (EE) of the squamous mucosa were performed and gastric fluid sampled after 24-28 hour feed removal. Seventy-two hours after EE, blood was collected at 1400 hours on pasture, following 12-hour feed removal, 60 minutes and 120 minutes after consuming 1 Kg of the pelleted diet. Data were submitted to ANOVA with sample time in a repeated statement. Mean number of crib-bites in 24 hours was 1,558±303 with CB peaking prior to and during the afternoon feeding (1530 hours). There were no differences in the number (Mann-Whitney U=36.5, P=0.59) and severity (Mann-Whitney U=36.0, P=0.54) of ulcers and prevalence of hyperkeratosis (Fisher's P=0.34) between CB and NCB. There was no difference (P=0.74) in gastric pH of CB compared to NCB (3.92 vs. 3.78±0.29). There was no effect of CB (P=0.17) on serum gastrin concentration (12.29 vs. 9.46±1.39 pg/ml). Feeding increased gastrin (P<0.01), but magnitude of the response was not different (P=0.14) between CB and NCB. Horses exhibiting CB did not have a higher prevalence of mucosal damage or altered gastrin response to feeding. These results suggest that gastrointestinal irritation is unlikely to be associated with CB in established cribbers maintained on pasture.

Brooding and selection on low mortality reduce fearfulness in domestic chicks

T.B. Rodenburg, K.A. Uitdehaag, E.D. Ellen and J. Komen, Wageningen University, Animal Breeding and Genomics Centre, P.O. Box 338, 6700 AH Wageningen, Netherlands

Fearfulness plays an important role in the development of feather pecking in laying hens: birds that are fearful at young age, have an increased risk to develop feather pecking later in life. Brooding by a mother hen and selective breeding may provide tools to reduce fearfulness and feather pecking. The aim of this study was to investigate the joint effect of brooding and selection on low mortality on fearfulness of domestic chicks. Birds in the experiment were either selected for low mortality in group housing and for egg production (low mortality line) or for egg production only (control line) for two generations. These lines originated from the same population. Twenty groups of 10 birds from each line were used. Within each line, ten groups were brooded by a foster mother and ten groups were non-brooded. At five to six weeks of age, the chicks were tested in an open field test for five minutes. The latencies to vocalise, stand up and walk and the number of distress calls and steps were recorded. Data were analysed using ANOVA, testing effects of line, mother and their interaction (with group as experimental unit). Brooded chicks had a shorter latency to stand up ($F_{1,33}=5.34$; $P<0.05$) and to walk ($F_{1,33}=8.20$; $P<0.01$) in the open field and walked more ($F_{1,33}=14.19$; $P<0.001$) compared with non-brooded chicks. Furthermore, chicks from the low mortality line uttered more distress calls ($F_{1,33}=6.50$; $P<0.05$) and walked more ($F_{1,33}=5.07$; $P<0.05$) in the open field than chicks from the control line. No significant interaction effects between line and brooding were found. These results show that both selection on low mortality and brooding reduce fearfulness in domestic chicks, and may reduce the propensity to develop feather pecking later in life.

Selection for low or high social motivation affects inter-individual discrimination in young and adult Japanese quail (*Coturnix japonica*)

C. Schweitzer, P. Poindron and C. Arnould, I.N.R.A., P.R.C., UMR INRA 85 / CNRS 6175 / Université de Tours / Haras Nationaux, 37380 Nouzilly, France

Rearing birds in large groups can induce the emergence of agonistic behaviours and social disturbances during encounters with strangers. A better knowledge of individual discrimination in birds is needed to understand and manage the consequences of these confrontations. Therefore we investigated inter-individual discrimination in quail after 24h of pair contact and how social motivation and age may affect this discrimination. A two-choice test was used to investigate discrimination between a familiar and an unfamiliar conspecific in young (one week) and adult (six weeks) Japanese quail selected for a high (HSR) or low (LSR) social reinstatement behaviour (n=32/line/age). At one week, the discrimination expressed by a preferential proximity and interaction with the familiar bird was evidenced in LSR quail (time spent and number of entries into the close zone: $p<0.05$, pecks: $p<0.01$, Wilcoxon's signed rank test) but not in HSR quail ($p>0.40$ in all cases). Either HSR chicks were able to discriminate but they would not express a preference because of their high social motivation, or they may have developed the ability to discriminate later than LSR chicks. At six weeks, male HSR quail displayed individual discrimination by preferentially directing social interactions toward the stranger (pecks: $p<0.05$, threats: $p<0.01$). Similar tendencies, although not significant, were observed in LSR quail ($p<0.10$ in both cases). A higher aggressive response towards the stranger in HSR quail could explain this difference. To verify this, aggressive behaviour of six-week-old male LSR (n=26) and HSR (n=20) quail was assessed in a separate experiment, by measuring their pecking activity towards a stranger presented behind a clear screen. Indeed, pecking activity was highest in HSR quail ($p<0.01$, Mann-Whitney U test), thus supporting our hypothesis. In conclusion, this study reveals for the first time in this species, the influence of social motivation and age, on the display of individual discrimination.

The effect of selective breeding for fitness traits on aggression during feeding in dairy cows

J.M. Gibbons, R.D. Donald, R.M. Turl, L.A. Maggs, A.B. Lawrence and M.J. Haskell, Animal Behaviour and Welfare, Sustainable Livestock Systems, SAC, West Mains Road, Edinburgh, EH9 3JG, United Kingdom

There is a general consensus within the UK dairy industry that health and longevity of dairy cows is declining. This may, in part, be due to breeding goals that focus primarily on production. There is an urgent need to address these concerns by developing balanced breeding goals to include 'fitness' traits (health, fertility and lifespan). However, it is important to consider any consequences that such breeding goals may have on dairy cow temperament and welfare. One aspect of dairy cow temperament that is particularly important to study is aggressiveness. Stress caused by aggressive interactions can negatively affect behaviour and feed intake, and cause some cows to alter their feeding times to avoid aggressive interactions. It is important to determine whether there maybe unforeseen consequences of breeding on levels of aggression. Aggressive behaviour was recorded on 402 first lactation Holstein-Friesian dairy cows selected from sires that scored high (H, n=229) and low (L, n=173) for 'fitness' traits to produce two treatment groups on 33 commercial farms. There were 6.93 + 3.76 H and 5.24 + 3.83 L (mean + S.D) cows per farm. Continuous focal sampling was used to record aggressive behaviour during feeding of the H and L cows within the herd. Cows from the H group were involved in more aggressive interactions (W_1=3.82 P=0.049), initiated more aggression (W_1=4.51 P=0.034) and received more aggression (W_1=4.36 P=0.037) than cows from the L group. There was a strong influence of management factors influencing aggression such as the quality of stockmanship, feedface design and nutrition. In conclusion, daughters from sires scoring high for fitness traits may be expressing a greater ability to maintain position at the feedface during an aggressive interaction. This highlights the importance of assessing the correlated effects of selective breeding, in this case for fitness, on behavioural traits.

Genetic validation of a method for measuring pig aggressiveness for use in selective breeding

S.P. Turner[1], R. Roehe[1], R.B. D'Eath[1], M.J. Farnworth[1,2], S.H. Ison[1], M. Farish[1], M.C. Jack[1] and A.B. Lawrence[1], [1]Scottish Agricultural College, Sustainable Livestock Systems, Bush Estate, Penicuik, Edinburgh EH26 0PH, United Kingdom, [2]Present address: Unitec, School of Natural Sciences, Carrington Road, Mount Albert, Private bag 92025 Auckland, New Zealand

Aggression following mixing of pigs is common but cannot be greatly reduced by low-cost management changes. Individual variability in aggression has been reported. This study estimated the genetic contribution to individual aggressiveness and, through genetic correlations, validated a method of predicting involvement in aggressive behaviour based on a count of skin lesions (lesion count, LC) accumulated following mixing as a rapidly quantifiable approach that could be used in selection programmes. The study used 1660 grower stage Yorkshire and Yorkshire X Landrace pigs fed ad libitum and housed in solid-floored pens. The heritability of LC to three body areas (front, middle and rear of the body) was 0.21-0.26. Two behavioural traits were found to have a moderate to high h^2 similar to that of growth traits; duration of involvement in reciprocal fighting ($h^2=0.43\pm0.04$) and delivery of non-reciprocated bullying ($h^2=0.31\pm0.04$), whilst receipt of bullying had a lower h^2 ($h^2=0.08\pm0.03$). Genetic correlations (r_g) suggested that lesions to the front region of the body were associated with reciprocal fighting ($r_g=0.67\pm0.04$) and receipt of bullying ($r_g=0.70\pm0.11$) whilst those to the middle and rear were associated primarily with receipt of bullying ($r_g=0.80\pm0.05, 0.79\pm0.05$). Pigs which engaged in reciprocal fighting bullied other animals ($r_g=0.84\pm0.04$) but rarely received bullying themselves ($r_g=-0.41\pm0.14$). Positive correlations between LC 24h and 3 weeks after mixing were found, especially for lesions to the middle and rear of the body, indicating that post-mixing lesions are predictive of those received under stable group conditions. A genetic merit index using lesions to the front, middle and rear regions as separate traits should allow selection against animals involved in post-mixing fighting and the delivery of bullying. Long-term impacts on injuries from more chronic aggression should also result from selection on LC.

Neonatal piglet mortality in relation to housing system and breeding value for piglet survival rate

L.J. Pedersen[1], G.H.M. Jørgensen[2] and I.L. Andersen[2], [1]Faculty of Agricultural Sciences, Århus University, Dept. of Animal Health, Welfare and Nutrition, P.O. Box 50, 8830 Tjele, Denmark, [2]Norwegian University of Life Sciences, Dept. of Animal and Aquacultural Sciences, P.O. Box 50003, N-1432 Ås, Norway

The present study investigated the interaction between genotype and housing system on causes of neonatal piglet mortality. Eighty primiparous landrace x yorkshire sows were selected, of which half had a high breeding value for survival rate (live piglets day 5/total born piglets) until day 5 (H) and the other half had a low breeding value for survival rate until day 5 (L). Sows were randomly selected within breeding group to farrow in crates (C) or pens (P). Data on individual piglets were collected at birth and included amongst other inter-birth-interval (IBI), birth order, weight at birth, rectal temperature and early suckling behaviour. All dead piglets were autopsied. Causes of mortality was divided into stillborn, starvation, crushed, disease and other causes. Potential risk factors of dying were estimated using a GLIMMIXED model. There were no significant effect of housing on the risk of a piglet to be stillborn ($F_{1,73}$=0.1, NS), be crushed ($F_{1,53}$=1.4, NS) or the risk of piglets to die due to starvation ($F_{1,53}$=0.3, NS). There tended to be a higher risk to be stillborn in litters of low compared to high breeding value ($F_{1,73=}$3.4, P=0.06). The risk of being stillborn were also higher in piglets with a long crown-rump measure ($F_{1,937}$=20, P=0.002), in piglets born late in the birth order ($F_{1,937}$=30, P<0.0001) and in piglets born after a long IBI ($F_{1,937}$=7.6, P=0.006). The lower the weight of the piglets the higher were the risk to be crushed ($F_{1,1050}$=18, P<0.0001) and to die of starvation ($F_{1,1050}$=19, P<0.0001). In addition, the lower the drop in rectal temperature 2h after birth the higher risk of being crushed ($F_{1,1050}$=4.6, P=0.03), die due to starvation ($F_{1,1050}$=16.6, P<0.0001) and die due to diseases ($F_{1,1050}$=4.9, P=0.03). The results emphasise the importance of piglets traits related to thermoregulatory ability for the survival of the newborn piglets.

Brain gene expression differences related to fearfulness in red junglefowl

M. Jöngren and P. Jensen, IFM Biology, Division of Zoology, Linköping University, 58183 Linköping, Sweden

The biology of fear is central to animal welfare and is a major target for domestication changes. We studied fear responses of birds from two different captive populations of red jungle fowl (RJF) originating from different zoos and known to differ in fearfulness. A total of 31 males and 32 females were tested in a ground predator test, an aerial predator test, and for tonic immobility. In the predator tests, the behaviour of each bird was quantified for five minutes before and after a simulated predator exposure, and in the tonic immobility test, we measured numbers of induction attempts and times to first move, and to righting. Correlations between 2-6 variables from each test were entered into a non-rotated factor analysis, extracting one major fearfulness factor (explaining 27% of the variance among the females and 21% of the males). The two most and the two least fearful birds within each sex and population were then selected based on factor scores (fearful average score: 1.41±0.26, non-fearful average score: -1.35±0.21; T-test: P<0.001). From each of these birds, the midbrain region (Nucleus tracus solitarii, Medulla oblongata, Pons, Mesencephalon, Diencephalon and the Pituitary) was collected and gene expression compared between groups using a 14k chicken cDNA microarray. Thirteen significantly differentially expressed (DE) genes (M>1, B>0) were found between the fearful and non-fearful females, but none between males. Among the DE genes, we identified the neuroprotein Axin1, two potential DNA/RNA regulating proteins and an unknown transcript situated in a well studied QTL region on chromosome 1, known to affect several domestication related traits. There were no significant gene expression differences between the two populations. The differentially expressed genes may be part of a possible molecular mechanism controlling fear responses in fowl.

Associative learning by cattle to enable welfare-friendly virtual fences

C. Lee[1], J.M. Henshall[1], T.J. Wark[2], C.C. Crossman[2], M.T. Reed[1], H.G. Brewer[1] and A.D. Fisher[1],
[1]CSIRO, Livestock Industries, Locked Bag 1, Armidale NSW, Australia, [2]CSIRO, ICT Centre, PO Box 883, Kenmore QLD 4069, Australia

An ability of cattle to readily associate a non-aversive cue with an aversive but non-noxious electric shock should enable virtual fences to control cattle in an ethical manner similar to conventional electric fencing. This 3-week study examined whether cattle location can be controlled by an audio conditioning stimulus. Audio (784 Hz tone) and shock (600V, 250 mW) cues were delivered by remote control to collars on heifers to prevent access to an exclusion zone (EZ). Weeks 1-2 tested the heifers' learning of the association between the audio conditioning stimulus and the shock response stimulus. In week 3, the effect of dispensing with the conditioning stimulus was tested. Heifers were randomly allocated to two treatments (n=12). Treatment 1 received an audio conditioning stimulus and a shock response stimulus, as in the first 2 weeks. Treatment 2 received no conditioning stimulus and a shock response stimulus on EZ entry. Data were analysed by logistic regression. On average, heifers received 2 shocks per training session in week 1 and 1 shock in week 2. There was a significant difference (P<0.001) in the ratios of behaviours shown in response to both the audio and shock cues between weeks 1 and 2 with more heifers turning in response to the audio cue in week 2 (24) than in week 1 (7). When the virtual fence was moved in week 2, 80% of animals ignored the first audio stimulus, but the proportion failing to respond to the second audio dropped to 46%, indicating that animals had learned the association between the audio and shock cues. In week 3, heifers received significantly fewer shock cues (2.6±0.6 vs. 7.1±1.1) when a conditioning stimulus was used (P=0.001). This study demonstrates that the use of an audio cue is an effective conditioning stimulus for virtual fencing of cattle.

The use by pigs of information from a mirror

D.M. Broom and H. Sena, Centre for Animal Welfare and Anthrozoology, Department of Veterinary Medicine, University of Cambridge, Madingley Road, Cambridge CB3 0ES, United Kingdom

Animals of a few species have been demonstrated to learn that what is seen in a mirror is not behind the mirror but is reflected in it, could pigs do this? Eight pigs were given three hours of experience with a mirror in a novel pen. They spent time looking at the reflections in the mirror. After this, the time spent in different parts of the pen, including behind the mirror and behind a 3m. long solid barrier put at right angles to and abutting the centre of the mirror, were recorded. Each pig was then put in a small pen from which it could see the mirror and a food bowl reflected in it but behind the barrier so not directly visible to the pig. When released, each pig approached the mirror. Seven out of eight pigs then walked away from the mirror around and behind the end of the barrier away from the mirror to get the food. A fan prevented detection of the food position by odour. Time spent behind the barrier was greater than expected from the control study. The mirror was then replaced by wire mesh and six out of eight pigs went behind the wire mesh to get to the food. In a further control, the food bowl behind the wire mesh was empty but the food bowl behind the barrier had food in it. The pig went behind the wire mesh to look for the food. These results indicate that pigs can be aware that a mirror indicates what is in a particular position in front of the mirror rather than what is behind it.

Can pigs learn a new motor action by observation?

S. Held[1,2], S. Cunningham[1], S. Jones[1], E. Murphy[1], M. Friel[1], R.W. Byrne[2] and M.T. Mendl[1], [1]university of bristol, department of clinical veterinary science, langford, bristol bs40 5du, United Kingdom, [2]School of Psychology, The University of St Andrews, St Andrews, Fife, KY16 9JP, United Kingdom

Pigs can learn by observation about a food location, but whether they can also learn new behaviours by observation is unknown. Knowledge of the ability to copy may help in the design of animal-operated equipment, and shed light on the transmission of damaging behaviours such as tail-biting. We therefore investigated whether pigs can learn from observation how to move a lever for food. Six test pigs observed a demonstrator move a lever to the left; six other test pigs observed a demonstrator move it to the right. The lever consisted of a cross attached to a rod. All demonstrators pushed the cross with their snouts/cheeks. For 12 control pigs the lever moved to the left or right without a demonstrator. After four demonstration sessions, pigs were tested for the first time. Thereafter every test was preceded by a demonstration session. All lever movements were rewarded, and all pigs completed 36 tests. Test pigs did not move the lever in the demonstrated direction more often than controls, but used the cross more often (Mann-Whitney tests, N=12; % movements in demonstrated direction - test pigs: median=48.61, first quartile=0.0, third quartile=96.5; control pigs: median=47.22, first quartile=1.39, third quartile=91.0; W=154.0, p>0.05; % movements using cross - test pigs: 95.83,55.56,100.00; control pigs: 13.90,0.69,29.90; W=195.5, p<0.01). Nine test pigs but only five control pigs used the cross in their first test (Chi-square=2.743, df=1, p<0.1). Test pigs were also less likely than controls to use their shoulders/necks (% movements using shoulder/neck - test pigs: 0.0,0.0,2.085; control pigs: 4.17,0,32.64; W=112.5, p<0.05). The test animals' propensity to use the demonstrated lever part and action indicates for the first time at least some level of response facilitation in the domestic pig. The lack of preference for the demonstrated directions, however, suggests that pigs lack the ability to fully copy a new motor action.

A spatial judgement task to determine emotional state in non-human animals

O. Burman, R. Parker, E. Paul and M. Mendl, University of Bristol, Langford House, Langford, Bristol, BS40 5DU, United Kingdom

Humans experiencing different emotional states display mood-congruent cognitive biases. For example, people in a negative emotional state make negative judgements about ambiguous stimuli. We have proposed that such biases may be used as indicators of animal emotional state. Here, we use a spatial judgement task, in which animals are trained to expect food in one location and not another, to determine whether rats in relatively positive or negative emotional states respond differently to ambiguous stimuli of intermediate spatial location. We housed 24 rats with environmental enrichment for seven weeks. Enrichment was removed for half the animals before training ('U': unenriched), but left in place for the remaining rats ('E': enriched). We predicted that 'U' animals, being in an assumed negative emotional state, would judge ambiguous locations negatively compared to 'E' animals. After six training days, the rats successfully discriminated between the rewarded and unrewarded locations in terms of an increased latency(s) to arrive at the unrewarded location (repeated measures GLM: $F_{1,6}=34.22$, P=0.001, rewarded 12.1±0.8s; unrewarded 19.8±1.9s), with no housing treatment difference ($F_{1,6}=2.2$, P=0.189, 'E' 17.8±1.7s; 'U' 14.1±1.1s). The subjects then received three days of testing in which three ambiguous 'probe' locations, intermediate between the rewarded and unrewarded locations, were introduced (with a probability 0.23). A significant Probe*Treatment interaction ($F_{2,4}=7.16$, P=0.048) revealed that whilst there was no difference between the treatments in the rats' judgement of two out of the three probe locations, the exception was when the ambiguous probe was positioned closest to the unrewarded location, with 'U' rats running slower than 'E' rats ('U' 14.7±4.5s; 'E' 6.7±0.6s). This result supports our prediction that 'U' rats, in an assumed relatively negative emotional state, show a negative judgement bias and are more likely to respond to ambiguous stimuli as if they predicted a negative event (i.e. no food) than 'E' rats.

Hunger in pregnant sows; the effects of fibre type and number of daily feedings

M.B. Jensen and L.J. Pedersen, University of Aarhus, Faculty of Agricultural Sciences, Department of Animal Health, Welfare and Nutrition, Blichers Allé 20 - P.O. Box 50, 8830 Tjele, Denmark

Fibrous diets may reduce hunger in pregnant sows. However, the satiating properties of different types of fibres, including their ability to maintain satiety over time, are unclear. The aim was to assess the effect of fibre type and number of daily feedings on the level of hunger. Four blocks of 15 pregnant sows were allocated to one of five diets within blocks. Four diets were offered restrictively (22 MJ net energy/day): a control diet (wheat/barley; 17.5% fibre), and three fibrous diets formulated to include 35% fibre by including pectin feed, potato pulp, or sugar beat pulp. The fifth diet was a mixture of the three fibrous diets, offered ad libitum. The experimental period included 2 periods of 4 weeks each. Sows were fed once daily (0800 hours) during the first period and twice daily (0800 and 1500 hours) during the second period, or vice versa. Towards the end of each period hunger was assessed using operant conditioning (progressive ratio; PR5 modified; 27 g food reinforcement). Sows were tested in a balanced design at 0700, 0900, 1200 and 1900 hours. The data were analysed using variance component analysis. No differences were found between the four diets offered restrictively, but ad libitum fed sows had a lower maximum level of responding (85, 79, 74, 84, 28 (SE 6.1); for control, pectin, potato, beat, and ad libitum; $P<0.001$). Among the restrictively fed sows one daily feeding vs. two daily feedings resulted in a higher maximum level of responding at 1900 hours (86 vs. 74 (SE 5.6); $P<0.01$), but not at 0700 (91), 0900 (68) and 1200 hours (83). The lower level of responding at 1900 hours in sows fed twice daily illustrates that allocating the daily feed allowance in two portions results in a better satiation during the night than one daily feeding.

Implications of food patch distribution on social foraging behaviour

L.R. Thomsen[1,2] and B.L. Nielsen[2], [1]University of Southern Denmark, Insitute of Biology, Campusvej 55, DK-5230 Odense M, Denmark, [2]University of Aarhus, Institute of Animal Health, Welfare and Nutrition, Blichers Allé 20, DK-8830 Tjele, Denmark

Spatial distribution of resources is an important factor influencing feeding behaviour of social animals. The aim of the study was to investigate how feeding behaviour of slaughter pigs changes when distance between feeding sites is increased. Groups of four slaughter pigs (n=8, 35.4±2.7 kg) were tested in feeding arenas daily during 12 days. Each group was exposed four times to three tests in random order. In tests four buckets containing food pellets would be placed either "close" (15-25cm), "medium" (245-290cm) or "far" (460cm) apart in a rectangular pattern. Each group was allowed to eat for approx. 20 min. Tests were video recorded and number of feeding bouts per pig, length of feeding bouts per pig, relative time spent eating per pig and number of agonistic encounters per group was registered. Food rations were increased during the test period corresponding with pig growth, and relative food intake per group was estimated by weighing refusals. Data were analysed using a general linear mixed model (SAS). There was an effect of bucket position on all variables (intake: F=11.74, p<0.0001; time spent eating: F=6.44, p=0.0029; no. of bouts: F=41.11, p<0.0001; bout length: F=5.67, p=0.0055; no. of agonistic encounters: F=12.32, p=0.0008). In close position there were more bouts (close=29.1, medium=21.6, far=19.3), higher food intake (close=86.8%, medium=78.0%, far=74.4%) and more agonistic encounters (close=33.0, medium=22.9, far=21.6) than in medium and far positions, but no difference between medium and far positions. For time spent eating only close and far positions differed (close=19.5%, medium=18.2%, far=16.8%). For bout length results were less clear with longer bouts in close position than in medium position, but otherwise no difference (close=37.0sec, medium=50.4sec, far=46.9sec). We conclude that feeding behaviour is markedly different when resources are clumped and even a small separation of troughs alters the feeding pattern and level of aggression profoundly.

Effects of feed restriction and social isolation on the time budgets and growth rate of female grower pigs

B.H. Stevens[1], P.H. Hemsworth[1], J.L. Barnett[2] and A.J. Tilbrook[3], [1]The University of Melbourne, Animal Welfare Science Centre, Faculty of Land and Food Resources, The University of Melbourne, Victoria 3010, Australia, [2]The Department of Primary Industries, Animal Welfare Science Centre, 600 Sneydes Rd, Werribee, Victoria, 3030, Australia, [3]Monash University, Department of Physiology, Building 13F, Monash University, Victoria, 3800, Australia

Assessment of welfare remains a contentious issue in the scientific and wider community. This study examined the effects on pigs of two mild stressors imposed over 6 weeks. A 2x2 factorial design was used to examine two main effects: social contact, (1) unrestricted social contact – individual housing with visual, tactile, auditory and olfactory contact with pigs and (2) restricted social contact – individual housing with only olfactory and auditory contact with pigs; feed level, (1) ad libitum and (2) 70% ad libitum feed intake. A total of 90 female pigs in three time replicates entered treatments at 11 weeks of age and remained in treatment for up to 6 weeks. Within each replicate, animals were allocated to one of the four treatments and catheterised at either 0, 2, or 4 weeks of treatment for serial blood sampling. Pigs were weighed fortnightly and video cameras used to continuously record behaviour. To examine time budgets of behaviour, video footage from one day in each of weeks 4-6 was point sampled every 15 minutes from 0930-1430h recording posture and behaviours (interactions with: floor; pen; neighbouring pig; feeder; drinker or none of these behaviours). Analysis of variance indicated no main effects on weight gain in the first 2 weeks of treatment, but feed restricted pigs gained less weight from 2-4weeks (p<0.002) and 4-6weeks (p<0.008) than ad libitum fed pigs. There were no main effects on the proportion of times the pigs were observed in each posture, or performing most behaviours. However, feed restricted pigs were observed to interact with the feeder a significantly higher proportion of the time (p<0.021) than ad libitum fed pigs. Social isolation had no effect on growth and behaviour, and while feed restriction reduced growth as expected, feed restricted pigs interacted more with the feeder.

Effects of dietary fermentable starch on behaviour of growing pigs in barren housing or on straw bedding

M. Oostindjer[1], J.E. Bolhuis[1], A.C. Bartels[1], H. Van Den Brand[1] and W.J.J. Gerrits[2], [1]Wageningen University, Wageningen Institute of Animal Sciences, Adaptation Physiology Group, Marijkeweg 40, 6700 AH Wageningen, Netherlands, [2]Wageningen University, Wageningen Institute of Animal Sciences, Animal Nutrition Group, Marijkeweg 40, 6700 AH Wageningen, Netherlands

Dietary fermentable carbohydrates may affect behaviour of restrictedly-fed pigs. Fermentation of carbohydrates results in a gradual, prolonged energy supply and may therefore promote inter-meal satiety and welfare. We investigated the effects of fermentable carbohydrates (native potato starch) on behaviour of growing pigs in barren housing (B) and on straw bedding (S). In a 2x2 factorial arrangement, 32 groups of eight pigs were assigned to either pregelatinised (P) or native (N) potato starch (35%) and to B or S housing. Pigs were fed two daily meals at 2.5 times energy requirement for maintenance. Behaviour was recorded at d21 for five hours distributed over the day, using 2-min instantaneous scan sampling. Activity over the 24-h cycle was scored in week 5 using 9-min instantaneous scan sampling. Data were analysed using GLM with group as experimental unit. S-housed pigs on the P-diet were more active than S-housed pigs on the N-diet, particularly during the light period (63.4±1.4 vs. 55.4±1.1% of time), whereas effects of starch type on activity were less obvious in B-housing (45.9±1.5 vs. 44.0±1.3% during light period; starch type x housing interaction, $P=0.05$), possibly because P-fed pigs in B-housing, unlike their S-housed counterparts, lacked a suitable substrate for their increased foraging motivation. No other starch type x housing interactions were found, however. In the morning, housing effects, such as more exploration and less oral manipulation of pen mates and fixtures in S-housed as compared to B-housed pigs, overshadowed those of diet ($P<0.05$). Before the second meal in the afternoon, however, P-fed pigs showed more locomotion, aggression ($P<0.05$) and oral manipulation of pen mates ($P<0.10$). After the second meal, P-fed pigs showed more activity, exploratory behaviour ($P<0.05$) and manipulation of pen fixtures than N-fed pigs ($P<0.10$). These results likely reflect a lower inter-meal satiety in P-fed pigs. In conclusion, dietary fermentable starch may enhance interprandial satiety and improve welfare in restrictedly-fed pigs.

Maturational changes in the prevalence of behavioural problems in guide and service dogs

D.L. Duffy and J.A. Serpell, University of Pennsylvania School of Veterinary Medicine, Department of Clinical Studies, 3900 Delancey Street, Philadelphia, PA, 19104-6010, USA

A standardised behavioural survey (C-BARQ™) was used to study canine behavioural development in five distinct populations of guide/service dogs (sample sizes ranged from 253 to 1,167). The survey was completed by puppy-raisers when dogs were 6 and 12 months old. Wilcoxon Signed Rank tests revealed consistent patterns of behavioural change occurring between these two time points. Where they occurred, age-related changes in C-BARQ factor and item scores were remarkably consistent across the 5 different organisations. Most problematical behaviours showed improvement with age: e.g. "chewing on inappropriate objects" ($P<0.001$, all organisations), "nervousness on stairs" ($P<0.05$, all organisations), "house soiling" ($P<0.005$, all organisations), "tail chasing" ($P<0.001$, all organisations), "separation-related problems" ($P<0.05$, 4 organisations), "owner-directed aggression" ($P<0.025$, 4 organisations) and "trainability" ($P<0.001$, 4 organisations). However, a few behaviours grew worse: e.g. "stranger-directed aggression" ($P<0.05$, all organisations), "attachment/attention-seeking" ($P<0.001$, 4 organisations), "excitability" ($P<0.001$, 4 organisations) and "escaping/roaming" ($P<0.025$, 3 organisations). Only one behaviour – "persistent barking" – showed opposite patterns of change between two organisations ($P<0.025$ at each school in opposite directions), a result that could be attributed to differences in breed composition between these two populations of dogs (N=226 German Shepherds (GS), 145 Golden Retrievers (GR), 210 Labrador Retrievers (LR) and 160 Lab/Golden crosses (LGX) at one school versus N=54 GS, 64 GR, 1,029 LR and 18 LGX at the other). Consistencies across organisations in the direction of change for many behaviours suggest that such effects are maturational and to some degree independent of environmental influences. However, further study of individual differences in rearing environment may still shed useful light on the reasons why some dogs deviate from their expected developmental trajectories.

Does a sheep's motivation to avoid hot conditions correspond to the physiological cost of remaining in those conditions?

A.D. Fisher[1], N. Roberts[1,2], L.R. Matthews[3] and G.N. Hinch[2], [1]CSIRO Livestock Industries, Locked Bag 1, Armidale NSW 2350, Australia, [2]University of New England, University of New England, Armidale NSW 2351, Australia, [3]AgResearch, Animal Behaviour & Welfare Research Centre, Private Bag 3123, Hamilton 3240, New Zealand

Two experiments investigated whether an animal's motivation to avoid hot conditions was aligned with its physiological response to the same conditions. In Experiment 1 (physiology), Merino ewes (n=12 per treatment) were assigned to 5 d of heat challenge: 20, 25, 30 or 35 °C with 70% humidity. In Experiment 2, 6 Merino ewes were trained to press a panel to open a door between the heat challenge pen and a cool pen at 20 °C. Each animal was exposed to the same temperatures as Experiment 1 and four fixed work ratios (1, 4, 10 and 25 panel presses), with one temperature and fixed ratio per 9-h test session. In Experiment 1, following ANOVA, respiration rates for the 20, 25, 30 or 35 °C treatments were 100±4.4, 132±3.5, 184±2.6 and 216±5.4 min^{-1}, respectively ($F_{3,43}$=58.92; P<0.001), and animal body temperatures were 39.3±0.03, 39.4±0.01, 39.8±0.01 and 39.9±0.06 °C, respectively ($F_{3,42}$=10.58; P<0.001). These results indicate that animals increased their adaptive responses as the temperature increased to 25 °C and above, but the treatments of 30 °C and particularly 35 °C represented a physiological challenge to the animals, resulting in increased body temperature. In Experiment 2, following REML analysis, the number of cool pen entries per session increased with increasing temperature, with 8.2, 11.3, 13.9 and 17.8 entries for the 20, 25, 30 or 35 °C treatments, respectively (pooled s.e.=0.92; Wald$_{3,87}$=187; P<0.001). The number of rewards obtained did not differ between fixed ratios of 1, 4 and 10, but declined at a fixed ratio of 25 (P<0.05). Similarly, animal respiration rates only increased at a fixed ratio of 25 (P<0.05). In conclusion, animals were motivated to access the cool pen in response to physiologically challenging temperatures, but at high access cost, a combination of moderate work for cool pen access and increasing their respiration rates was adopted by sheep to maintain thermoregulation.

Does age affect social preferences in farmed silver fox females (*Vulpes vulpes*)?

A.K. Akre, M. Bakken and A.L. Hovland, Norwegian University of Life Sciences, Department of Animal and Aquacultural Sciences, P.O. Box 5003, 1432 Ås, Norway

Social contact may improve welfare in farmed animals provided animals perceive this contact as more positive than being housed solitary. *Vulpes vulpes* are socially flexible animals and the black colour morph, the silver fox, may establish social groups in captivity. However, social contact is usually limited to just after weaning. The aim of this study was to investigate social preferences in young silver fox females (cubs and juveniles) and their motives for seeking social contact. Forty-two foxes, including fourteen test animals, were used in this study. At 9½ and 24½ weeks of age, the females attended a three-choice preference test. The preference cages (240×180×75cm) were made of wire mesh and consisted of one test cage (240×80cm) and three visually separated stimuli cages (80×100cm) holding a sister, an unfamiliar female and an empty cage. The test animal could interact with the stimulus foxes through the wire mesh walls. The test fox' position and behaviour were recorded using instantaneous sampling every tenth-minute for twenty-six hours. Data were analysed using Van der Waerden and Wilcoxon signed-rank test. There was a clear preference for seeking contact with a conspecific at 9½ weeks, spending 91.9±4.41% of time in front of the stimulus cubs ($p<0.01$). There were no preference towards familiar versus unfamiliar cubs ($p>0.05$), however they had a tendency to play more in front of the unfamiliar cub ($p=0.067$). When juvenile, there was no preference for seeking contact or being solitary ($p>0.05$), though the test female directed significantly more agonistic signals towards the unfamiliar female ($p<0.01$). In conclusion, our results indicate that there is a change in silver fox females' social preference with age. The motives for seeking contact as cubs were non-aggressive and possibly related to play motivation, whereas the aggressive behaviour displayed towards the unfamiliar females in juveniles indicate an increased competitive motivation.

Pet rabbit housing and welfare in the Netherlands

P. Koene, B. Beerda and F. Schepers, Wageningen University, Animal Sciences, Marijkeweg 40, 6709 PG Wageningen, Netherlands

There is no research showing the behaviour and welfare of the one million pet rabbits in the Netherlands. The aim of this study was to initiate research on behaviour and welfare of pet rabbits in relation to housing conditions in Dutch households. A survey was placed on the internet with questions about rabbit housing systems, general care given to rabbits and behaviour of rabbits. The answers of 919 respondents were analysed. Behaviour observations of 90 rabbits were done at people's homes. The rabbits were observed during a contact test (contacts made by the rabbit with the observer's hand) and a handling test (resistance against the observer lifting the rabbit). Rabbits were also observed for an hour in their home cage. The survey revealed that the average lifespan of the rabbits is three years (maximal lifespan is eight to twelve years). About half of the respondents housed their rabbit(s) solitary, while the majority housed them in a small cage. Solitary housed rabbits made more contacts (9.8) than rabbits housed together (2.1; M-W, $p<0.001$). Rabbits in a small housing system made more contacts (12.7) than rabbits in a large housing system (4.1; M-W, $p<0.01$). Nearly one fourth of the rabbits showed strong resistance when lifted up, indicating socialisation problems. The observations in the home cage indicated that solitary rabbits showed stereotypical behaviour. Rabbits living in large cages or living outdoors showed a more natural time budget than rabbits living in small cages of living indoors (M-W, $P<0.01$). In conclusion, the results of the survey and of the behavioural tests and observations showed the presence of abnormal behaviour, effects of cage size on time budgets and large discrepancies with a natural time budget of rabbits. This research indicates the existence of major behaviour and welfare problems in pet rabbits in the Netherlands.

Prenatal stress amplifies the pain associated with tail-docking in piglets

K.M.D. Rutherford[1], S.K. Robson[1], R.D. Donald[1], S. Jarvis[1], D.A. Sandercock[2], E.M. Scott[3], A.M. Nolan[2] and A.B. Lawrence[1], [1]SAC, Animal Behaviour & Welfare, West Mains Road, Edinburgh, EH9 3JG, United Kingdom, [2]University of Glasgow, Institute of Comparative Medicine, Glasgow, G61 1QH, United Kingdom, [3]University of Glasgow, Department of Statistics, Glasgow, G12 8QW, United Kingdom

Various deleterious effects of prenatal stress are well established in many species including pigs. However, the impact of prenatal stress on neonatal responses to potentially painful injury has not been widely examined. This study examined whether mixing of pregnant gilts with older sows, which has been previously shown to alter the stress reactivity of offspring piglets, also altered the behavioural response to tail-docking as a putative indication of the pain associated with this procedure. Maternal stress was induced in gilts (n=16) during the second third of pregnancy by mixing with older sows. Control gilts (n=11) were kept in undisturbed social groups throughout pregnancy. Two- or three-day old piglets from stressed mothers or controls were observed immediately following either tail-docking (Stressed: n=94, Control: n=51) or a sham procedure (Stressed: n=75, Control: n=51) involving a similar degree of handling. Behavioural responses following docking or handling only were quantified for one minute. Piglet behaviour was altered by tail-docking. A pain score based on the duration of abnormal tail states and frequency of tail wagging was found to very accurately predict the docked status of piglets (sensitivity =98.9%, specificity =94.7%). REML was used to compare scores with prenatal stress status, sex, and weight fitted as fixed effects and litter fitted as a random effect. Offspring from stressed mothers had higher pain scores following tail-docking, compared to offspring from control mothers (Prenatal stress: 37.5±1.8, Control: 25.3±2.3; Wald =4.92, p=0.027). There was no effect of piglet sex (Wald=0.02, p=0.88) or body weight (Wald =0.03, p=0.85) on the behavioural response to docking. Short-term behavioural observation following tail-docking allows the severity of piglets' acute pain response to be well quantified. These data provide some evidence that the experience of stress in utero can result in a heightened response to injury in early life.

Weaning behaviour and weight gain in dairy calves suckling the dam in an automatic milking system or milk-fed from an automatic feeder

S. Fröberg[1], I. Olsson[1], L. Lidfors[2] and K. Svennersten-Sjaunja[1], [1]Swedish university of agricultural sciences, Animal nutrition and management, Kungsängens research centre, SE-753 23 Uppsala, Sweden, [2]Swedish university of agricultural sciences, Animal environment and health, P.O.Box 234, SE-532 23 Skara, Sweden

The aim was to investigate if calves of Swedish Red Breed suckling the dams freely (FS; n=18) in an automatic milking (AM) system would have more behavioural signs of stress and less growth rate after abrupt weaning at 8 weeks of age, than groups of 10-12 calves given a low (LM; n=23) or high milk (HM; n=23) substitute allowance from an automatic teat-feeder (DeLaval). The daily allowances corresponded to the energy content of 5 and 9 kg, respectively, whole milk. All treatments had concentrate and hay ad libitum. At weaning FS calves were moved to a pen similar to the other treatments. The weight gain was recorded until 10 weeks. The behaviours of focal calves were observed 24-22 h before and 0-2, 10-12, 24-26, 72-74 h post-weaning, instantaneously at 3-minute intervals (lie, eat concentrate, ruminate) or by frequency within every third minute (eat hay, vocalise, move). Behaviours which were normally and approximately Poisson distributed and growth were analysed using the MIXED, GENMOD and GLM procedures, respectively (SAS version 9.1). After weaning FS calves were 'lying' less and 'moving' more (p<0.001) compared to LM and HM calves, until 72 h post-weaning. FS calves also had less recordings of 'eat concentrate', 'ruminate' (p<0.001), and 'eat hay', and more recordings of 'vocalise' (p<0.01). More HM (76%) than LM (39%) and FS (30%) calves (p<0.05) displayed 'cross-sucking'. Numerically more LM (22%) and HM (53%) than FS (8%) calves exhibited 'tongue-rolling'. LM and HM calves had a higher post-weaning weight gain (p<0.001), but the advantage in pre-weaning weight of FS calves remained until week 10. In conclusion, free suckling in an AM-system compared to automatic teat-feeding in group housing resulted in more behavioural signs of stress during weaning until 72 h post-weaning and lower post-weaning weight gain.

Abrupt and gradual weaning increases cross-sucking in dairy calves on an automated milk feeder

A.M. De Passille[1], B. Sweeney[2] and J. Rushen[1], [1]Agriculture and Agri-Food Canada, PO 1000 6947 Highway 7, Agassiz, BC V0M 1A0, Canada, [2]University of Edinburgh, Edinburgh, Scotland, United Kingdom

Cross-sucking is low when calves drink large volumes of milk, but occurs when the calves are weaned off milk. We examined the effect of gradual weaning on cross-sucking. Twenty-eight Holstein calves were housed in groups of 4, and fed 12L/d of milk from an automated milk feeder, with weaning from milk completed by d41. They were assigned to one of three treatments. ABRUPT calves were weaned abruptly off milk on d41. The milk ration of EARLY GRADUAL calves was reduced by 0.5L/d beginning on d19. The milk ration of LATE GRADUAL calves was reduced by 2.5L/d beginning on d37. Treatments were applied within groups. All occurrences of cross-sucking were observed on d18, d30, d39, d41, d42 and d49. Over the whole period, the daily duration of cross-sucking was higher (analysis of variance: $p<0.05$) in the EARLY GRADUAL calves (8.9±1.6min) than the ABRUPT calves (5.3±1.4min), with the LATE GRADUAL calves intermediate between the two (6.8±1.4min). This was due to a difference in the frequency of bouts of cross-sucking ($p=0.02$) with no difference in the bout duration ($p=0.67$). Prior to final weaning, the EARLY GRADUAL calves showed a greater daily duration of cross-sucking (6.3±0.8min) than the LATE GRADUAL calves (1.0±0.8min) and the ABRUPT calves (1.2±0.8min)($p<0.001$). After final weaning, there was no difference ($p=0.34$) in the daily duration of cross-sucking (EARLY GRADUAL:11.7±2.8min.; LATE GRADUAL: 12.7±2.7min.; ABRUPT: 9.4±2.6min.). Among ABRUPT calves, the daily duration of cross-sucking did not change with age before weaning, but increased after weaning ($p=0.001$). For calves fed with an automated milk feeder, weaning off milk provokes cross-sucking. Gradual weaning increases the amount of cross-sucking that occurs prior to weaning. Delaying the beginning of gradual weaning reduces this effect. Although gradual weaning increases grain consumption prior to weaning and reduces weight loss at weaning, it does not reduce cross-sucking.

Physiological and behavioural alterations in disbudded kids with and without local anaesthesia using lidocaine 2% (L2%)

R.A. Nava, A. Ramírez, E. Ramírez, J. Gutiérrez and L. Alvarez, Facultad de Medicina Veterinaria y Zootecnia, UNAM, Producción Animal, Tequisquiapan, Querétaro, México, 76750, Tequisquiapan, Querétaro, México, Mexico

In order to evaluate the physiological and behavioural response in disbudded kids with or without local anaesthesia, two experiments were conducted. In the first experiment, 4 groups were randomly formed: Anaesthesia (A, n=12), kids receiving 2ml of L2% around each horn bud 20 minutes before disbudding with thermal cauterisation; Saline (Sa, n=11), kids receiving 2ml of Sa before disbudding; Simulation (S, n=11), the procedure was only simulated; and Control (C, n=11), kids disbudded without any previous treatment. Plasmatic cortisol levels, respiratory and cardiac frequency were measured from 20 minutes before up to 4 hours after disbudding; frequency and intensity of kicks and vocalisations during the procedure were also measured. In the second experiment, to elucidate the real effect of injecting the anaesthetic, plasmatic cortisol was determined in 13 kids injected with 2ml of L2% around each horn bud (7 disbudded, 6 control). A repeated measures Anova, Kruskall-Wallis, Mann-Whitney and the Fisher test were used. Disbudding caused a significant increase in cortisol levels that lasted 2 hours. Disbudded kids had higher levels of cortisol ($p<0.05$) than S during 1 hour after disbudding. Respiratory and cardiac frequencies were not significantly affected by treatment. In groups A, Sa and C, the percentage of animals showing high intensity behaviours (kicks: 83, 72 and 100%; vocalisations: 83, 81 and 100% respectively) were greater than in S (0 and 9% respectively; $p<0.05$). Injecting the anaesthetic induced a slight increase in cortisol levels by itself; when the injection was followed by disbudding, the cortisol elevation was significantly higher (71.3 ± 16.8 vs. 115.4 ± 15.5, nmol/L, $p<0.05$) and more extended, suggesting that anaesthesia was not effective. It is concluded that disbudding by thermal cauterisation induces an acute cortisol elevation and increases the presentation of behaviours indicative of discomfort and pain. L2% around each horn bud did not inhibit these alterations.

The effect of group size and personality on social foraging: the distribution of sheep across patches

P. Michelena, A.M. Sibbald, H.W. Erhard and J.E. McLeod, Macaulay Institute, Ecology Group, Craigiebuckler, Aberdeen, AB15 8QH, United Kingdom

Group cohesion in social animals foraging in patchy environments is likely to be affected by the interaction between intraspecific competition and attraction. We investigated this in an experiment with sheep, using different group sizes and personality types. Twenty 'shy' and twenty 'bold' sheep were identified from an indoor exploration test, on the basis of having a low propensity ('shy') or a high propensity ('bold') to explore at a distance from conspecifics. Four weeks later, groups of 2, 4, 6 or 8 bold or shy sheep were introduced successively into 45 x 5m grass arenas, with one 5 x 5m patch of preferred vegetation at each end, for 30-minute periods. Five replications were carried out, with all group sizes tested each time for each personality type. Sheep grazed on or close to these patches, but seldom more than 4-5 individuals at a time, suggesting that crowding might be affecting foraging at the highest densities. The smallest groups grazed together on the same patch, but there was an increasing likelihood of splitting into subgroups as group size increased, with equal-sized subgroups most commonly grazing the two patches simultaneously. This effect was greatest in bold sheep, which tended to split into subgroups at smaller group sizes than shy sheep. Our spatial analysis confirmed that the tendency to split into subgroups increased with group size ($F_{1,24}=28.9$, $P<0.001$) and that this happened sooner in bold sheep than in shy ones ($F_{1,24}=5.1$, $P=0.03$). These results suggest that both the sensitivity to crowding and the propensity to move away from conspecifics underlie such group fusion-fission dynamics. This study provides new insight into the mechanisms by which group-living herbivores distribute themselves across patchy resources in a way that minimises interference competition and demonstrates the importance of individual variability for spatial organisation at the level of the group.

Patterns of decision-making in group movements of grazing sheep

A. Ramseyer[1,2], B. Thierry[2], A. Boissy[1] and B. Dumont[1], [1]INRA, UR 1213 Herbivores, 63122 Saint Genès Champanelle, France, Metropolitan, [2]CNRS, IPHC DEPE, UMR 7178, 23 rue Becquerel, 67087 Strasbourg, France, Metropolitan

Movements of grazing herbivores are usually driven by a limited number of individual animals called leaders. To investigate the mechanisms underlying group movements in grazing sheep (Ovis aries), we observed recruitment processes during the 30 minutes preceding departure (i.e. start of the first individual from a group that was resting) and animal order in the following movement, in a flock of 19 ewe-lambs. Once their social relationships were established, animals were continuously videotaped in a 5-ha field by two observers who recorded 102 group movements from a resting position. Videos were analysed using the Observer software. The frequency of head movements and the number of individuals standing-up in the direction of the movement significantly increased in the last two minutes before departure (Friedman test, $P=0.006$, $P<0.001$). These behaviours were interpreted as promoting group recruitment (multiple regression analysis, $F=11.8$, $R^2=0.51$, $P<0.001$). The first individual to move recruited more mates when it had a greater number of neighbours within one meter and when it departed from the group edge (multiple regression analysis, $F=7.0$, $R^2=0.25$, $P<0.001$). Departure and movement orders were relatively stable (Spearman test, $r_s=0.64$, $P<0.01$ and $r_s=0.56$, $P<0.05$), with one ewe-lamb being more often seen as the first mover ($Chi^2=6.8$, $P<0.01$), and another one more often taking the front position during movements ($Chi^2=7.1$, $P<0.01$). The analysis did not provide any evidence of the influence of dominance status on the order of group movements. In contrast, sheep linked by affiliative bonds were usually observed in successive positions during group movements. We conclude that leadership is distributed as several individuals influenced the decision-making process. These findings provide a first insight into the collective decisions of grazing sheep. They could have implications for the modification of grazing patterns of the whole flock by selecting leaders with desirable grazing behaviour.

Do sheep categorise plant species according to their botanical family?

C. Ginane and B. Dumont, INRA, UR1213- Herbivores Research Unit, Centre de Clermont-Ferrand / Theix, 63122 Saint Genès Champanelle, France, Metropolitan

When grazing on complex and botanically diverse pastures, herbivores may categorise food items in order to reduce the quantity of information to deal with for diet selection. Here, we studied if sheep are able to consider different botanical families (grasses: G vs. legumes: L) as distinct categories, and to generalise their knowledge about one plant species to those from the same family. Twenty-four ewe-lambs aged 6 months were assigned to four groups characterised by the plant species: tall fescue (G), perennial ryegrass (G), red clover (L) or sainfoin (L), against which they were negatively conditioned with mild doses of Lithium Chloride (LiCl, 70 mg/kg body weight). Animals preference between freshly cut cocksfoot (G) and alfalfa (L) was measured weekly after four successive conditionings occurring once in each of four one-week long periods (P1 to P4). We assumed that the preference for alfalfa would be lower for animals conditioned against other legume species than for those conditioned against grasses. We analysed the effect of the conditioned family (G or L), species nested in family (e.g. sainfoin(legume)), period and interactions, on intakes and preferences using the SAS mixed procedure with the repeated statement for period effect. Intake during conditioning decreased through periods (from 98 to 54 g DM, p=0.0004) indicating that lambs perceived the negative effect of LiCl. During choice tests, alfalfa preferences for animals conditioned against different species inside each family group were similar (p=0.37). On the contrary, alfalfa preference was lower for the L than the G groups over the whole experiment (0.65 vs. 0.78, p=0.04), without significant interaction between family and period (p=0.52). These results are the first experimental evidence that grazing sheep are able to categorise plant species on the basis of their botanical family.

Reshaping grazing behaviour through alternative paddock management system

A.L. Schmitt[1,2] and W.M. Murphy[1], [1]The University of Vermont, Plant and Soil Science, Hills Building, 05405, USA, [2]University of Santa Catariana, Centro de Ciencias Agrarias, 88040900, Brazil

Management methods have been shown to adjust grazing behaviour. Dairy cows can be expected to anticipate being moved to a fresh paddock after milking and synchronise their behaviour to this prediction. This results in reduction of grazing time and intake, especially around reward time. This experiment tested the effect of paddock management in reshaping behavioural patterns of cows to improve use of the pasture resource. Forty experimental cows were paired and allocated to ten sets of four animals (blocks) according to milk yield, lactation days, and lactation number. Within each set, cows were randomly allocated to two replicates of two treatments with 10 animals in each group. Two control groups of cows were under typical half-day paddock management. Two treatment groups were moved to paddocks with distinct internal design. Half-day paddocks were fenced into two areas: main (85% of area) and remainder (15% of area). The main area was made available to cows when they arrived in a paddock. The remaining area of fresh forage was offered to them during the waiting period, at 12 AM and 10 PM before next milking. Behavioural and production differences between treatment and control were tested using ANOVA. The magnitude of the difference was evaluated through Tukey-Kramer HSD. Treatment cows grazed 72 min d^{-1} more (498.5 min SD=46.8 vs. 426.5 min SD=48.5 P<0.01), and had shorter rumination (471.7 min SD=45.6 vs. 424.4 min SD=51.8 P<0.01) than controls and equal leisure daily times (277.8 min SD=66.7 vs. 251.4 a SD=62.8 P<0.12). These behavioural adjustments happened only when the fenced off pasture strip was released to the treatment group. A simple management practice such as this can greatly reshape cow grazing behaviour, improve farm overall productivity, and may affect profitability.

Value of space to group housed hens in cages; techniques for addressing behavioural requirements in commercial conditions

J.J. Cooper, S.E. Redgate, T. Luddem and M.J. Albentosa, University of Lincoln, Department of Biological Sciences, Riseholme Park, Lincoln, LN2 2LG, United Kingdom

This paper presents studies investigating the value of additional space to groups of laying hens in enriched cages. Between 2005 and 2012, conventional wire cages will be phased out and replaced by enriched cages for laying hens. Enriched cages should provide nest-boxes, perches and scratching areas and at least 750cm2 floor space per bird. Our project sought to investigate the value hens placed on additional cage space as well as develop the methods required to investigate resource value in group housed animals. We used four techniques; the first involving observation of hens at various stocking densities in enriched cages. There was high competition for resources such as perch and food trough at high stocking densities. Comfort activities such as wing/leg stretching were largely absent. The second technique involved moving hens from enriched cages to open pens, where locomotion and comfort activities were the immediate response to provision of additional space. In the third approach, groups of hens were allowed to choose between cages of various dimensions. In these studies, hens consistently distributed so as to maximise individual space irrespective of flock size or available space. Finally, hens were allowed to work to gain access to additional space. Hens would overcome a high cost to access additional space, however work rate was largely dependent on the size of the cage they were leaving, and not the size of the space they were accessing. These results suggest reducing stocking density reduces competition for resources in enriched cages, however activities such as wing-flapping and stretching are still restricted in enriched cages. Also at the stocking densities allowed in enriched cages, hens strive to increase individual space. The work suggests that using social costs such as competition represents a suitable means of assessing value of resources in group housed animals under commercial conditions.

Pre-laying behaviour of nest layers and litter layers

M. Zupan[1], A. Kruschwitz[2], T. Buchwalder[3] and B. Huber-Eicher[4], [1]Ministry of Agriculture, Forestry and Food, Dunajska 56-58, 1000 Ljubljana, Slovenia, [2]Tierseuchenkasse Baden-Württemberg, Schaflandstr. 3/3, 70736 Fellbach, Germany, [3]Centre for Proper housing of Poultry and Rabbits, Burgerweg 22, 3052 Zollikofen, Switzerland, [4]Swiss College of Agriculture, Länggasse 85, 3052 Zollikofen, Switzerland

According to the Swiss Animal Welfare Act nests for laying hens must be approved to assure the provision of nest sites that meet the hens' needs. The aim of the study was to verify whether there are different types of laying hens with regard to the nest choice within the white commercial hybrid (LSL). 24 individually housed hens had a choice of 2 nests: a tray with wood shavings (litter tray) and a tray with wood shavings plus a nest box (nest box). The behaviour 1 h prior to oviposition and the chosen nest site was recorded for the first 20 eggs laid. The majority of hens preferred the nest box, the rest preferred the litter tray. All hens laid exclusively in one nest site, thus consistency in nest site selection was very strong. Nest layers spent less time exploring than did litter layers ($F_{1,45}=5,22$; $P=0.04$) and stayed longer in the nest site when an egg was laid ($F_{1,45}=3,75$; $P=0.07$). Both types of layers entered their preferred nest site at the same frequency ($F_{1,45}=0.83$; $P=0.38$). To summarise, we provide evidence for two types of layers, laying either in closed nest sites or in open litter trays. They concomitantly performed different pre-laying behaviour, which points to two different behavioural strategies. If litter layers in our experiment correspond to floor layers in commercial poultry systems, this may mean that the latter lay on the floor not because they are prevented by other hens to lay in the nests but because they do not perceive the offered nests as optimal nest sites. Our conclusion is that to prevent floor eggs one should offer different types of nests that correspond to the needs of both types of layers.

Does the height of perches matter for laying hens?

L. Schrader, K. Krösmann and B. Müller, Friedrich-Loeffler-Institut, Institute of Animal Welfare and Animal Husbandry, Doernbergstrasse 25/27, 29223 Celle, Germany

During resting, laying hens show a strong priority for perches and they prefer high compared to low perches. However, it has never been tested whether the height of perches is important independent from the perch itself. In three trials a total of 36 groups with 9 laying hens balanced for two strains (Lohmann Brown, Lohmann Selected Leghorn) were kept in compartments (300x185 cm) in which they were offered two different resting areas each. Resting areas consisted of elements (90x65 cm) of two heights (L: 15 cm or H: 60 cm) with either a plastic grid (G) or perches (P) on top of the elements. Preferences for perches vs. grid were tested with the combinations HP/HG and LP/LG, preferences for high vs. low resting area with the combinations HP/LP and HG/LG, and preferences for the height vs. perch with the combination HG/LP. The combination HP/LG served as a control. In the 10th week after start of lay the number of hens on the resting areas was video-recorded and counted in 1h-intervals throughout two consecutive nights for 8 hours respectively. The LSQ-means of the differences between the proportions of hens on the different elements were estimated with an ANOVA, and t-values were used to test whether these differences were unequal 0. Hens of both strains significantly preferred the high vs. low elements (H-L: 73.7%, t=4.1, P<0.001) and the perches vs. grid (P-G: 38.2%, t=2.1, P<0.05) during night. In the control treatment the perches on high elements were preferred vs. the grid on low elements (HP-LG: 56.8%, t=2.8, P<0.01). Interestingly, when high elements with grid were offered in combination with low elements with perches, the hens significantly preferred the high elements with grid (HG-LP: 86.6%, t=4.3, P<0.001). The latter result suggests that in order to improve the welfare of laying hens not only should perches be offered but perches should be elevated.

Preliminary study on the use of perches by pullets of two different lines in a floor system with natural day-light

C. Keppler[1], E.F. Kaleta[2] and U. Knierim[1], [1]Faculty of Agricultural Sciences, Department of Farm Animal Behaviour and Husbandry, Nordbahnhofstr. 1a, 37213 Witzenhausen, Germany, [2]Veterinary Faculty, Clinic for Birds, Reptiles, Amphibians and Fish, Frankfurterstr.91-93, 35329 Giessen, Germany

Learning to use perches is important for laying hens in non-cage systems. We compared use of perches in 12 to 18 weeks old pullets of two laying strains (LSL, LT) under three different light-regimes (one natural and two artificial with different day-lengths) and two light intensities (higher: about 133lux vs. lower: 43lux). Twelve groups of each 100 non beak-trimmed pullets were kept in littered pens (8.8 animals/m²) with natural light and 11cm perches/animal. Mean percentages of perching pullets were recorded from videos with instantaneous-scan-sampling for 5 minutes every hour during daylight over 9 days. Perching at night was directly observed once on two days and, additionally, once eight weeks after moving to laying pens. Analyses of variance with fixed factors strain, light-regime and light intensity were carried out. At day-time, significantly more LSL-pullets were perching (LSL:22.8±3.5% vs. LT:16.4±2.2%, F=27.6, p=0.002). Light-regime had no influence (F=0.7, p=0.54), but contrary to LT, LSL pullets tended to perch more under higher light intensity (higher:25.7±2.0% vs. lower:19.9±0.8%, F=5.2, p=0.062, interaction: F=6.0, p=0.05). When 25% of the birds were on perches, it was obviously difficult for further hens to perch. During the night, more LSL pullets tended to perch (LSL:18.1±24.3% vs. LT:2.5±3.0%,F=5.1, p=0.065). There was a slight interaction between strain and light-regime (F=3.9, p=0.082). Contrary to LT, LSL perched more with natural light-regime (natural:44.2±27.93 vs. artificial I:9.5±6.82 and artificial II:0.6±0.35), probably due to a longer dawn. Although our results confirm reports from practice that LSL-pullets are more ready to perch, after moving to the laying pens both strains nearly completely perched at night (LSL:99.7±0.47%; LT:92.5%±17.8), though with greater variation in LT. However, LT had more floor-eggs (LT:8.8±5.5% vs. LSL:2.8±1.4%, F=6.7, n=6, p=0.027, ANOVA), suggesting that low use of perches during rearing partly impairs the later use of elevated structures in the laying house.

Locomotion ability and lying behaviour of dairy cows at pasture or in cubicle housing over a complete lactation

G. Olmos[1,2], L. Boyle[1], J.F. Mee[1] and A.J. Hanlon[2], [1]Teagasc, Moorepark, Dairy Production Research Centre, Fermoy, Co. Cork, Ireland, [2]University College Dublin, School of Agriculture, Food Science & Veterinary Medicine, Belfield, Dublin 4, Ireland

The aim of this study was to compare pasture and indoor cubicle environments for a full lactation using cow locomotion ability and lying times as indicators of cow welfare. Forty-six Holstein-Friesians (mean calving date February 28th 2007) were blocked and randomly allocated to a PASTURE or INDOOR environment. Locomotion ability was assessed every 2 weeks [0 to 220 days in milk (DIM)] using 6 gait aspects: spine curvature, speed, tracking, head carriage, abduction/adduction and general symmetry. Each aspect was scored from 1 (normal) to 5 (abnormal). The hazard of a cow reaching a threshold of 3 at day t post-calving, given that it had not reached the threshold by time t-1, was modelled using survival analysis. Lying times were recorded every 5min using voltage data-loggers (Tinytag Plus, UK) on 13 cows/group for 3x48h periods at 33, 83 and 193 DIM, respectively. Data were analysed using mixed models for repeated measures. INDOOR cows had a greater hazard ($P<0.01$) of scoring ≥3 for tracking, head carriage, general symmetry and abduction/adduction (hazard ratios: 2.5, 2.8, 2.7 and 3.6 respectively) and for spine curvature (2.1, $P=0.069$). Total lying times were shorter ($F=12.45$, $P<0.001$) for INDOOR compared to PASTURE cows (18.1h, s.e. 0.71 vs 20.5h, s.e. 0.73). INDOOR cows had more lying bouts ($F=10.22$, $P<0.05$) than PASTURE cows (22.8, s.e. 1.37 vs 16.3, s.e. 1.45). Cows housed INDOORS for a complete lactation had reduced locomotion ability and resting behaviour compared to cows at PASTURE. These differences help to explain the higher incidence of lameness in housed compared to pastured cows.

Trade-off decisions of dairy cow feeding preferences

F.C. Lang[1], D.J. Roberts[2], S.D. Healy[3] and M.J. Haskell[1], [1]Scottish Agricultural College, Sustainable Livestock Systems, Sir Stephen Watson Building, SAC, Bush Estate, Midlothian, Scotland, EH26 OPH, United Kingdom, [2]Scottish Agricultural College, Dairy Research, Midpark House, Bankend Road, SAC, Dumfries, Scotland, DG1 4SZ, United Kingdom, [3]The University of Edinburgh, Evolutionary Ecology of Animal Cognition, School of Biological Sciences, Kings Buildings, Edinburgh, EH9 3JT, United Kingdom

Maximising feed intake in dairy cows is essential for maintaining milk production and health; however, competition can cause increased aggression and displacements. Choice tests were used to determine how individuals perceive their feeding environment, with specific emphasis on difficulties faced by subordinates. Thirty Holstein Friesian cows were used in the study. Cows were moved individually from a group pen and walked freely towards the test pen containing a Y-maze. At the end of the Y-maze was either a black or white bin. Half of the cows were trained to make an association between the black bin and high quality food (HQF), and the white bin and low quality food (LQF). The other half were trained with the opposite combination. The status of each cow was assessed and cows were split into pairs consisting of a dominant and subordinate individual. Test 1 determined if cows had correctly learned the association between colour and food quality. Cows had to achieve an 80% success rate of HQF preference before they could proceed. Two further tests were presented. In Test 2, the subordinate cow was presented with two bins of HQF, one of which had a dominant cow feeding from it. In Test 3, cows had a choice of HQF or LQF, with a dominant cow at the HQF bin. Chi-squared tests were used to test for significant differences between treatments. Cows showed a significant preference for feeding on HQF alone rather than next to a dominant, $p<0.001$. When they were asked to trade-off feed quality with feeding next to a dominant, the majority chose to feed alone on LQF, $p<0.01$. These results suggest that social status within a herd could significantly affect feeding behaviour, especially in situations of high competition. This is important to consider for feed management and barrier design.

Rubber flooring in front of the feed bunk: effect of surface compressibility on restless behaviour and preferences in dairy cows

N. Krebs, S.L. Berry and C.B. Tucker, University of California-Davis, Animal Science, Meyer Hall, CA-95616, USA

Alternatives to concrete flooring improve dairy cattle heath and comfort. Dairy cows spend 4-6h/d feeding in front of the feeder. In addition, cows are often restrained for veterinarian care at the feeder. Little work has evaluated the importance of physical features such as compressibility of rubber flooring in this area. We compared three surfaces: concrete, a single layer or a double layer of rubber flooring (2x and 4x more compressible than concrete, respectively).The first phase of the experiment assessed restless behaviour while lactating cows were restrained on a single surface (n=16). Restless behaviour (number of steps) was measured for 30min/h for 4 h after morning milking. During the second or preference phase, cows had free access to pairs of treatments for 2 days (concrete+single-layer, concrete+double-layer, single+double-layer). Preference was assessed by the amount of time spent on each surface (4 h/24h) and the feed intake from each surface (over 24h). The amount of restless behaviour increased by 48% (P<0.001, Page's test) over the 4-h period. Cows were more likely to move their hooves when standing on either the single- (114.2 steps/h ±3.23) or double-layer (113.9±3.23) of rubber, compared to concrete (102.6±3.23, 0.05vs. 40.2%±6.55% of 4h after milking, P<0.05, t-test). These results indicate that cows preferred surfaces that are 4x more compressible than concrete. Cows showed less restless behaviour on concrete than on rubber surfaces, indicating that compressibility may influence their willingness to move. Restless behaviour did not differ between the single- and double-layer rubber flooring, indicating that factors other than compressibility, e.g. traction, may also affect restless behaviour.

Influence of feed barrier space allowance and design on behaviour of dairy cows

N.E. O'Connell, L.G. Baird and C.P. Ferris, Agri-Food and Biosciences Institute, Large Park, Hillsborough, Northern Ireland BT26 6DR, United Kingdom

This study examined whether silage feed barrier space allowance and design affected aggression at feeding, and ability of heifers to access the barrier. Fifteen animals (10 cows and 5 heifers) were each assigned in early lactation to one of four treatments: (1) 20 cm feed barrier space per cow, without head spaces; (2) 20 cm barrier space per cow, with head spaces; (3) 57 cm barrier space per cow, without head spaces; (4) 57 cm barrier space per cow, with head spaces. The animals were offered a total mixed ration of grass silage and concentrates during the study. A change-over design was used, comprising of four, 5-week periods. Behaviour was observed during weeks 4 and 5 of each period (except week 4 in Period 1). Each group was observed for 1 hour after morning and evening milking on 1 day in each observation week. Each observation was divided into 6 x 10-minute periods, and the frequency of aggressive behaviour at the feed barrier within each period was recorded. The proportion of heifers at the feed barrier was recorded from 30 minute scans over a 24 hour period during each recording week. Aggression data were analysed by Generalised Linear Mixed Model analysis, and proportion of heifers at the feed barrier was analysed by Analysis of Variance. Main and interactive effects of treatment factors, and interactions with time of day ('day' or 'night') were assessed. Aggressive behaviour increased when individual head spaces were provided at the low feed space allowance, but decreased when spaces were provided at the high feed space allowance (T1: 1.80, T2: 4.36, T3: 4.47, T4: 2.34 (freq/10 min), $P<0.01$). There were no significant treatment effects on the proportion of heifers at the feed barrier (average=0.33). There were no significant interactive effects between treatment factors and time period ($P>0.05$). The use of individual feed spaces appears to promote aggression at reduced feed barrier space allowances.

Animal experiments: environmental standardisation increases risk for spurious results

S.H. Richter[1], J.P. Garner[2] and H. Würbel[1], [1]University of Giessen, Animal Welfare and Ethology, Frankfurter Strasse 104, 35392 Giessen, Germany, [2]Purdue University, Animal Sciences, 125 South Russell Street, West Lafayette, IN 47907, USA

Subtle differences in laboratory or test conditions resulting in conflicting outcomes of animal experiments raised calls for more rigorous environmental standardisation. However, environmental standardisation reduces the external validity and could, therefore, cause, rather than cure, poor replicability of results from animal experiments. To test this hypothesis we used data from a multi-laboratory study on behavioural differences between three inbred mouse strains (n=144/strain) distributed across three laboratories, three independent batches per laboratory and two housing conditions (n=8 mice/strain/batch/housing condition). The resulting 18 cohorts, each standardised for a unique combination of laboratory, batch and housing condition, were treated as 18 replicates of a standardised experiment. These were compared with 18 replicates of a randomised experiment, each composed of 8 mice per strain that were pseudorandomly selected from 8 standardised replicates to mimic partial environmental variation. To compare the results of each replicate with "true" strain differences, we pooled all mice and calculated strain differences for five typical measures from each of four behavioural tests (elevated O-maze, open-field, novel-object, Morris water-maze). Between-replicate variance in mean strain differences across the 20 behavioural measures was significantly greater in standardised replicates (t-Test, $p<0.001$), indicating poorer replicability. To explore this further, we determined the proportion of false positive results, with false positives denoting significant strain differences in single replicates where there were no significant strain differences in the pooled data. The false positive rate was significantly lower in randomised replicates (Wilcoxon-Test, $p<0.05$), indicating greater specificity. Moreover, in contrast to randomised replicates, standardised replicates produced significantly more false positives than expected by chance (Binomial-Test, $p<0.001$). These findings demonstrate that environmental standardisation systematically increases the rate of spurious results and, therefore, generates a need for replicate studies. This fundamentally undermines its ethical goal to reduce animal use, in addition to the scientific and economic costs associated with it.

Feeding-place design affects agonistic behaviour in goats

N.M. Keil, J. Aschwanden, L. Gygax and B. Wechsler, Centre for proper housing of ruminants and pigs, Federal Veterinary Office, Tänikon, 8356 Ettenhausen, Switzerland

When feeding, a higher-ranking goat asserts her rank position by threatening or butting a lower-ranking goat if it comes too close. In two experiments with adult female goats, we tested whether and which type of modification at the feeding place can positively affect the feeding and agonistic behaviour. The experiments were conducted with 48 goat dyads stemming from eight groups (four each with/without horns and four each grouped as adults/juveniles according to a 2 x 2 factorial design). In experiment 1, the dyads fed side-by-side at two hayracks separated by four different types of partitions (short [50 cm] or long [110 cm], and either wire-mesh or solid wood), or without a partition. In experiment 2, the dyads had to share one hayrack, but one of the goats had the possibility to stand on a platform raised 80, 50, 25 or 0 cm above ground level. Simultaneous feeding time and frequency of agonistic interactions were each tested in a linear mixed-effects model for dependence on type of partition and height of platform. In experiment 1, the presence of any partition had strong positive effects on goat behaviour. Simultaneous feeding time ($19.5\pm2.4\%$ vs. $90.3\pm1.2\%$, all $p<0.001$) increased and the number of agonistic interactions/60s decreased (6.0 ± 0.54 vs. 1.4 ± 0.12, all $p<0.001$) compared to the situation without a partition. For horned goats and for dyads with a large rank-index difference, these effects were most obvious in the experimental conditions with a solid wood partition. In experiment 2, positive effects on feeding and agonistic behaviour were most consistent in the experimental condition with the platform at a height of 80 cm. In conclusion, partitions at the feeding place and feed access on different heights are effective in reducing competition between goats during feeding.

Species adequate housing and feeding for an opportunistic carnivore, the red fox (*Vulpes vulpes*)

C. Kistler[1], D. Hegglin[2], H. Würbel[3] and B. König[1], [1]University of Zurich, Institute of Zoology, Animal Behaviour, Winterthurerstrasse 190, CH-8057 Zurich, Switzerland, [2]SWILD - Urban Ecology & Wildlife Research, Wuhrstrasse 12, CH-8003 Zurich, Switzerland, [3]Justus Liebig University of Giessen, Institute of Veterinary Physiology, Division of Animal Welfare and Ethology, Frankfurter Strasse 104, D-35392 Giessen, Germany

Many zoos keep their animals in naturally looking enclosures, but whether the species' ecological and behavioural needs are adequately met is often unclear. We experimentally examined the effectiveness of (1) structural enrichment in modifying enclosure use, and of (2) feeding enrichment in eliciting appetitive behaviour in an opportunistic carnivore, the red fox (*Vulpes vulpes*). (1) Wooden structures simulated hedges or thicket in predetermined sectors. A group of four foxes was exposed to four consecutive treatments: structural enrichment in location 1, in location 2, removed and again in location 1. Sectors containing wooden structures were significantly preferred (Friedman-Test, Chi^2=8.1, p=0.036, df=3, n=4). (2) Four feeding procedures served as substitutes for natural food resources: Electronic feeders (EF), a self-service food box (SF), manual scattering of food (MS), an electronic dispenser (DIP). Treatments were offered consecutively in the following combinations: EF (unpredictable in time), EF and SF (unpredictable in time, plus time-consuming manipulation), EF and MS (unpredictable in time and space, plus time-consuming), and EF and DIP (highest degree of unpredictability in time and space, plus time consuming). The daily diet consisted of meat, nuts, and fruits. Compared to conventional feeding, where food was offered once daily, behavioural diversity (Chi^2=15.57, p=0.001, df=5, n=4), activity (Chi^2=16, p=0.001) and visibility (Chi^2=13.74, p=0.004) were significantly enhanced during the four enriched feeding procedures. The combination of EF and SF resulted in the highest behavioural diversity, whereas both EF and MS, and EF and DIP produced higher activity than EF and SF (post-hoc test after Conover, all p ≤ 0.05). Thus, red foxes benefit from a richly furnished enclosure and spatially and temporally variable feeding procedures.

Size and organisation of lying space have a large impact on resting pattern and social interactions in goats

K.E. Bøe and I.L. Andersen, Norwegian University of Life Sciences, Department of Animal- and Aquacultural Sciences, P.O. Box 5003, 1432 Aas, Norway

The aim of this experiment was to examine how the size and organisation of lying space affects resting pattern and social interactions in female milking goats. Twenty-four goats divided into 6 groups, were systematically rotated between six experimental pens (width x depth: 2 x 3 m) with lying areas of different size (small: 0.5 m², medium: 0.75m² or large: 1.0m²) and organisation (one vs. two floors). Resting pattern was analysed using instantaneous sampling with 10 minutes intervals for 24 hours, whereas social interactions were continuously observed for five hours between 09.00 and 14.00 during the last 24 hours of each experimental week. A mixed model ANOVA with size of resting area and lying space organisation as main factors was used to analyse the effect on resting and social behaviour. Group was specified as a random effect. The goats spent less time resting (P<0.01) and rested less simultaneously (P<0.001) when the lying area was small compared to a medium and large lying area. Time spent lying in the activity area also increased with decreasing lying space (P<0.01). The goats preferred resting close to a pen wall, and this occurred less often when the lying area was small (P<0.01). Resting in social contact with pen mates occurred in less than 6% of the observations lying, and this was not significantly affected by the size of the lying area. Social dominance had a strong impact on resting pattern. The amount of social interactions was not significantly affected by the size of the lying area, but there were significantly fewer displacements (P<0.01), and the overall aggression level was lower (P<0.05), when lying space was organised on two floors rather than one. In conclusion, time spent resting and resting pattern was more dependent on size (large, medium, small) than organisation (one vs. two floors) of the lying space, whereas this was the opposite for social interactions.

Mouse temperature preferences in laboratory housing with naturalistic nesting material

B.N. Gaskill, S.A. Rohr, E.A. Pajor, J.R. Lucas and J.P. Garner, Purdue University, Animal Sciences, 125 S. Russell Street, West Lafayette, IN 47907, USA

In laboratories, mice are housed at 20-24 °C which is below their thermoneutral zone (26-34 °C). We have shown that although mice prefer temperatures warmer than 20 °C, there is no suitable temperature for all animals or housing setups. We tested the hypothesis that providing mice nesting material will allow behavioural thermoregulation, and reduce preference against colder cages. We housed C57BL/6 mice (24 male and 24 female in same sex groups of 3) in standard laboratory conditions from 4-12 weeks of age. One cage of males and one cage of females were tested each week but only one sex received nesting material (8g). The sex receiving nesting material alternated weekly. Mice were tested in a set of 3 connected cages, each maintained at a different temperature using a water bath. For three days, mice were acclimated to one temperature (20 °C, 25 °C, or 30 °C) per day in a random order. Then mice were given access to all temperatures, and location, contact with nesting material, and behaviour (active, inactive, maintenance, nesting, and unknown) was recorded by instantaneous scan samples every 10 minutes from 72h of video, and time budgets calculated. Analysis used split-plot GLMs with post-hoc contrasts or Tukey tests. Nesting material affected temperature preference (GLM: $F_{2,144}=4.87$;p=0.009): mice preferred warmer temperatures, spending approximately 55% of their day at 30 °C, but nesting material doubled time spent in 20 °C from 7.6% to 15.6% (Tukey: p=0.002). Behaviour also differed according to contact (GLM:$F_{4,49}=104.1$;p<0.001): active behaviours were more likely to be observed when not in contact with nesting material (11.3% vs. 6.8%; Contrast:$F_{1,49}=7.30$;p=0.009), but inactive behaviours (48.2% vs. 0.81%; Contrast: $F_{1,49}=512.5$; p<0.001) and nesting behaviours (2.9% vs. 0.95%; Contrast:$F_{1,49}=8.68$;p=0.004) were more likely to be seen when in contact. We conclude that nesting material partially compensates for cooler temperatures, that nesting material is especially important when inactive, but mice still prefer warmer temperatures overall.

Personality factors affecting human-canine interactions

M. Wedl[1], B. Bauer[1], J. Dittami[1], I. Schöberl[1] and K. Kotrschal[1,2], [1]University of Vienna, Konrad-Lorenz-Research Station and Department of Behavioural Biology, Althanstraße 14, 1090 Vienna, Austria, [2]IEMT Austria, Margaretenstrasse 70, 1050 Vienna, Austria

The aim of our study is to investigate the relationship between personality, interactions and temporal structure of behaviours in human-dog dyads. We predict to see peculiar dyadic interaction styles and also, time patterning of interactions, which are influenced by the personalities of both partners. We started to explore these ideas in human-dog dyads. Twenty-three human-dog dyads participated in our study (11 male, 12 female owners, aged 23-68, with their medium and large sized intact male dogs, aged 1.5 to 6). During three sessions dyadic behaviours and interactions were observed and video-taped in different test situations (e.g. mild threatening of the dog by experimenter moving head and gazing at it; owner present or absent). Owners completed the NEO-FFI personality test and an attitude-toward-dog-scale. Dog personality data were obtained via observer rating and behavioural tests. Principal component analyses (PCA) served to condense dog personality ratings and the attitude-toward-dog-scale. The videos were behaviour coded with THE OBSERVER (Noldus) and the temporal structure of behaviours and interactions between dog and owner are analysed by THEME (Noldus). During the threatening situation with the owner present, male owners touched and held the dog for longer periods of time than female owners (Mann-Whitney-U: n=22, Z=-2.664, p=0.008 and Z=-1.988, p=0.047). The more conscientious (FFI axis 5) the owners, the shorter the dog barked and growled during this situation (Spearmans: n=22, r_s=-0.474, p=0.026 and r_s=-0.471, p=0.027); the more "inactive-uninterested" the dog (PCA axis 1) the longer the owner stroked the dog (r_s=0.587, p=0.004). The results suggest significant effects of owner gender and personality of both, owner and dog on the dyadic interactions. Temporal patterns of dyadic interactions (THEME analysis) are yet to be analysed and will also be addressed.

When ethologists, keepers and students listen to pig vocalisations: how do they evaluate the emotional state of the animals?

C. Tallet[1], M. Špinka[1] and I. Maruščáková[2], [1]Institute of Animal Science, Ethology Department, Pratelstvi 815, 10400 Prague 10, Czech Republic, [2]Faculty of Natural Sciences, Charles University, Department of Zoology, Viničná 7, 12844 Prague 2, Czech Republic

The evaluation of the emotional state of animals is essential to assess their welfare. Vocalisations are a good tool for this evaluation as they vary for instance according to the intensity of the emotion (high/low) and its valence (displeasure/pleasure) and they are easily measured. This study investigated the human perception of pig vocalisations, pigs being highly vocally expressive. We tested if humans discriminate different types of domestic piglet vocalisations on the basis of the intensity and the valence of the emotion expressed and if the experience with pigs influences this discrimination. Forty-eight records from two negative (castration, isolation) and two positive (reunion with the sow, after suckling) situations were played back to 36 ethologists studying pigs, 31 pig-keepers and 54 inexperienced students. Responders judged the emotional intensity of three records of each situation and the valence of three others on five-point scales (very low to very high intensity, high displeasure to high pleasure). Data were analysed with ANOVA and Scheffe post-hoc tests. Responders discriminated the situations through both intensity ($F_{3/44.8}=54$; $p<0.001$) and valence ($F_{3/44.1}=97$; $p<0.001$). Castration sounds were highly intensive (4.4 ± 0.1 vs 2.4 ± 0.1; $p<0,001$) and unpleasant (1.4 ± 0.1 vs 3.5 ± 0.1; $p<0,001$) compared to the other sounds. Reunion sounded more pleasant than isolation (3.7 ± 0.1 vs 3.2 ± 0.1, $p<0.01$). The identity of the record in each situation explained 25% of the variability for the intensity and 14% for the valence ($p<0.001$). The overall intensity was lower for the keepers than for the others ($F_{2/116}=9.4$; $p<0.001$), and the valence lower for the ethologists ($F_{2/116}=10$; $p<0.001$). We showed that humans can properly evaluate the emotions expressed in pigs' vocalisations, specifically their valence, probably based on particular characteristics of each record. Listening to pigs is thus useful to evaluate their welfare. Experience affects the sensitivity maybe due to different concerns about pigs breeding and habituation in keepers.

Handling during routine care affects responses of laboratory rats in standard behaviour tests

S. Cloutier, J. Panksepp and R.C. Newberry, Washington State University, Department of Veterinary and Comparative Anatomy, Pharmacology and Physiology, Center for the Study, P.O. Box 646520 Pullman WA, 99164-6520, USA

We hypothesised that the type of handling used by caretakers during routine care of laboratory rats affects rats' fear of humans and their subsequent responses in behaviour tests used in biomedical research. We assessed caretaker handling effects on male Sprague-Dawley rats in four treatments (N=8 rats/treatment) between 57 and 74 days of age: (1) Standard (control): handled on cage cleaning days (2x/week) only (picked up by tail and placed in clean cage); (2) Passive Hand: exposed to a motionless hand for 2 min daily; (3) Pinned: restrained on back for 2 min daily; (4) Tickled: tickled for 2 min daily. Behaviour was assessed between 78 and 80 days of age in a Human Approach test [2 min without hand (HA), followed by 3 min with a familiar (HAF) or unfamiliar (HAU) passive hand]. Ultrasonic vocalisations (50-kHz, interpreted as evidence for positive affective states) were recorded during each test. Tickled rats reared more frequently than Pinned rats in the HA (8±0.7 versus 5±0.6 rears), with other treatments intermediate [$F(3, 24)=3.44$, $P=0.03$]. Tickled rats emitted more 50-kHz calls during HAF (53±18.3 calls) than rats from the Standard and Passive Hand treatments (mean 7±1.7 calls), with Pinned rats intermediate [GLM, $F(3, 24)=4.49$, $P=0.01$]. They also spent more time near the hand in the HAF (110±11.6 s), and contacted it more frequently (3±0.9 contacts), than rats from the Standard and Passive Hand treatments (mean 62±9.9 s; 1±0.3 contacts), with Pinned rats intermediate [$F(3, 24)≥4.39$, $P≤0.01$]. In HAU, trends were similar but non-significant. The results suggest that incorporating playful handling (i.e., tickling) into the routine care of laboratory rats reduces fear of humans and can alter responses in behaviour tests.

Donkeys and bribes…maybe more than just a cartoon!

A. McLean, C.R. Heleski and L. Bauson, Michigan State University, Animal Science, Anthony Hall, East Lansing, 48824 Michigan, USA

There are approximately 54 million donkeys worldwide yet little research has been devoted to them. Anecdotally, handlers are frequently observed using harsh treatment to get donkeys to cross novel surfaces. We examined the application of learning theory to donkeys at a Michigan facility. The objective was to record their ability to complete a novel, potentially frightening task - crossing a tarpaulin - and compare different methods for accomplishing this. A pre-trial assessed the donkeys' behaviour using two methods, negative reinforcement (NR) (n=5) and NR + positive reinforcement (PR) (n=5). NR: pressure was applied via halter and lead rope; when donkey stepped forward, pressure was released. PR: donkeys were given a handful of grain for each step taken forward (NR + PR). However, unlike results from a previous horse study, PR did not appear successful with the donkeys. We proceeded at the pre-trial to test NR + luring (NR+L; n=3), which appeared more successful. L: grain was held in front of the donkey prior to subject stepping forward; subject received grain when stepping forward. Untested donkeys (n=36) were then assigned to either NR or NR+L. Each donkey was timed for 10 min and if they did not cross, they were considered to fail. The results: 4/18 in NR failed; 5/18 NR+L failed (not different, X^2=0.23, P=0.63). The first time the donkeys crossed the tarpaulin, time varied greatly but means did not differ (36.7±13.65 sec for NR; 41.7±14.51 sec for NR+L; F=0.07; P=0.80). A post hoc treatment was added immediately post-failure to encourage failures to cross. Plastic sheeting attached to a longe whip was waved behind donkeys in this treatment. Only 2 did not cross after this addition. This study provides evidence that harsh treatment should not be required to get donkeys to cross novel surfaces.

Maternal impact on beef calves' responses to humans

X. Boivin[1], F. Gilard[1] and D. Egal[2], [1]INRA, UR1213H-ACS, centre de recherche INRA de theix, 63122 St-Genés Champanelle, France, [2]INRA, UE expérimentale des monts d'Auvergne, INRA Le Roc, 63210 Orcival, France

Early life stages are important in building cattle-human relationships. Recent studies suggested also a strong influence of the maternal environment and could support traditional husbandry systems with daily long intervals of separation (e.g. twice a day suckling practices). The objectives of the present experiment were to evaluate the importance of the duration of early dam-calves separation and of initial human contact on calves' later responses to human and to investigate the relationship between responses of dams and calves. Thirty-three Salers calves (from two to four days of age) reared outdoors were involved in three-week treatments balanced according to sex and birth dates. T1 (n=11): 8h of separation from the dam per day and 5 min of individual stroking. T2 (n=11): One hour of separation and same human contact as T1 calves. T3 (n=11) same separation as T2 without stroking. At 3, 15 and 40 weeks of age, calves were tested in a standard arena test (AT): alone (2min) with a stationary (5min) and touching (2 min) human successively. At 15 and 45 weeks, they were also tested with the standard Docility test (DT: test of restraint). Dams were also tested with DT, two months before calving. Mann-Whitney tests and Spearman correlations were used for data analysis. No significant effect of duration of separation was observed. At week 3 but not later, T1 and T2 calves accepted more the human contact (AT: touching) than T3 ($P<0.01$). In AT (touching) and DT at the different ages, the behaviour towards humans and DT score of T1 and T2 but not of T3 calves were correlated with the dams' DT score (up to 0.7, $P<0.01$). Tactile contact in early age but not the duration of the separation from the dam seem beneficial for the human-animal relationship, but only for calves who have a docile dam.

Can we find an alternative to the electric prod for loading slaughter weight pigs?

S. Torrey[1], J.A. Correa[2], N. Devillers[1], H. Gonyou[3] and L. Faucitano[1], [1]Agriculture and Agri-Food Canada, Dairy and Swine Research and Development Centre, PO Box 90, 2000 College Street East, Sherbrooke, Quebec J1M1Z3, Canada, [2]Université Laval, Département des Sciences Animales, Pavillon Paul-Comtois, Québec, Québec G1K 7P4, Canada, [3]Prairie Swine Centre, Box 21057, 2105 – 8th Street East, Saskatoon, Saskatchewan, S7H 5N9, Canada

Although there is mounting evidence regarding the negative welfare implications of electric prod use for slaughter weight pigs, prods continue to be used, especially during loading when multiple pigs are moved through tight spaces. The objective of this study was to compare the behaviour of pigs at a commercial facility during loading with alternative moving devices (paddle and compressed air prod) to those loaded with an electric prod. Over three weeks, 270 pigs (BW=120±7 kg) were moved in cohorts of 5 through a corridor and up a 10° ramp onto a truck. Pigs were moved by one trained handler using one of three devices: electric prod (E), paddle (P) or compressed air prod (A). The time to load each group was recorded, and behaviour during loading was observed from video recordings. Data were analysed using the mixed model procedure in SAS. Time to load was different among the three moving devices ($P<0.001$; E:19.1±2.0 sec, P:39.4±5.0 sec, A:53.1±6.5 sec). With E, there were more slips and falls (sf) ($P<0.001$; E:1.8±0.2 sf/load, P:0.7±0.2 sf/load, A:0.3±0.2 sf/load) and overlaps ($P=0.04$; E:0.7±0.1 overlaps/load, P:0.3±0.1 overlaps/load, A:0.2±0.1 overlaps/load) than when pigs were loaded with A or P. When loaded with A, pigs turned around ($P<0.01$; E:0.5±0.2 turns/load, P:0.4±0.2 turns/load, A:1.3±0.2 turns/load) and backed up ($P<0.001$; E:0.4±0.4 backups/load, P:1.7±0.4 backups/load, A:3.1±0.4 backups/load) more often than pigs loaded with E or P. With A and P, pigs balked ($P<0.001$; E:0.8±0.3 balks/load, P:3.1±0.3 balks/load, A:3.3±0.3 balks/load) more often than pigs loaded with E. In conclusion, the electric prod moved pigs quickly but it is an aversive method with pigs exhibiting behaviour that may lead to injury. While the paddle appears to be a suitable substitute for the electric prod, the compressed air prod is not an acceptable alternative since loading took longer and involved more escape behaviour.

The relationships between handling prior to slaughter and stress in sheep and cattle at abattoirs

P.H. Hemsworth[1], M. Rice[1], L. Calleja[1], G.A. Karlen[1], G.J. Coleman[2] and J.L. Barnett[1], [1]University of Melbourne, Animal Welfare Science Centre, School of Agriculture and Food Systems, Gratton St, 3010 Victoria, Australia, [2]Monash University, Animal Welfare Science Centre, School of Psychology, Psychiatry & Psychological Medicine, Building 17, 3800 Clayton, Australia

This study examined handling-animal stress relationships at 2 sheep and 2 cattle abattoirs. The handling of 200 animals, of similar age from one property, was studied on 1 day at each abattoir. A bout criterion interval of 5 s was used to record the frequency of tactile, auditory and visual interactions by stockpeople directed to each animal from the forcing pen through to the stunning area. A total of 14 and 13 stockpeople handled the study sheep and cattle, respectively. The use of dogs on each sheep was scored based on duration and intensity of dog use. The head position of each animal, scored as head down or head up, was also recorded in response to both a stockperson handling animals and a stationary researcher located just prior to stunning. The number of vocalisations by cattle at the end of the single file race was also recorded using a bout criterion interval of 5 s. Blood samples for cortisol analysis were collected within 1 minute of slaughter. Linear regression was used to examine the relationships between these variables. The model that best predicted cortisol concentrations in sheep post-slaughter included the significant variables time of day, abattoir, dog use, head position of animal to the researcher and stockperson auditory interactions received (adjusted R square=0.21, $F_{6,371}$=48.5, P=0.000, using backward method). The model that best predicted cortisol concentrations in cattle included the significant variables abattoir, head position of animal to the stockperson, vocalisations by the animal and auditory interactions and goads received (adjusted R square=0.438, $F_{5,325}$=18.0, P=0.000, using backward method). In these equations, head down, increased dog use, reduced auditory interactions received and increased goads received were significantly associated with increased cortisol. These relationships suggest that training programs, targeting key stockperson behaviours, may reduce fear and stress in sheep and cattle at abattoirs.

Are responses of beef cattle to a human approaching the feed-barrier related to ease of handling?

I. Windschnurer[1], S. Waiblinger[1], P. Boulesteix[2] and X. Boivin[3], [1]Institute of Animal Husbandry and Welfare, University of Veterinary Medicine, Veterinärplatz 1, 1210 Vienna, Austria, [2]France Limousin Sélection, Pôle de Lanaud, 87220 Boisseuil, France, [3]UR1213H-ACS, INRA de Clermont-Ferrand/Theix, 63122 Saint-Genès Champanelle, France

Ease of handling of beef bulls is linked with work efficiency, safety, animal productivity and welfare. The docility test (DOC), a handling procedure restraining animals in the corner of a pen, has been extensively used in research but is also regularly applied to promising French Limousine beef sires in the selection process. Our objective was to evaluate, if a quicker and safer test assessing bulls' responses towards a human approaching at the feeding place (ADF), would be repeatable and related to responses in DOC. At the bull testing station of Lanaud, bulls aged 8–10 months were tested over 8 days. Animals were subjected thrice to ADF (ADF1 to 3) restrained in the head-lock feed-barrier, recording individual avoidance distances in steps of 10 cm (by person A) and filming more detailed behaviour. The avoidance distance was the distance at which the approaching human provoked a first avoidance movement (pulling back or turning the head more than 45°). Moreover, the docility test was performed once, recording a 7-rank docility note in 0.5 steps ranging from 1 (docile) to 4 (aggressive) (by either person B or C as experimenter, while the other person acted as observer). DOC was performed in-between ADF1 and ADF2. Due to practical constraints, some animals were tested only once with ADF. Spearman correlations and Mann-Whitney test were used for data analysis. Preliminary results suggest moderate test-retest repeatability of ADF (r_s=0.54, 0.62, 0.49, P<0.001, n=137, 120, 119, for ADF1/ADF2, ADF1/ADF3, ADF2/ADF3). Between ADF 1 and the mean docility score of persons B and C, a significant correlation was found (r_s=0.34, P<0.001, n=194), whereas for ADF 2 and ADF 3 only tendencies were found (P<0.10). The more time individual animals spent withdrawing maximally (stopped by the feed-barrier) during ADF, the higher were their avoidance distances and docility scores (r_s=0.41/0.37, P<0.001, n=178/179). Aggressive animals (n=13) withdrew longer than animals in all other categories (Z=-3.23, P=0.001, n=179). Our first results are promising, suggesting that ADF is to some extent a repeatable and valid measure of beef bulls' responses to humans.

Effects of sex, twinning and periconceptional ewe nutrition on lamb behavioural reactions to separation from the ewe

C.E. Hernandez[1,2,3], M.H. Oliver[1,2], F.H. Bloomfield[1,2], L.R. Matthews[2,3] and J.E. Harding[1,2], [1]Liggins Institute, University of Auckland, Private Bag 92019, Auckland, New Zealand, [2]National Research Centre for Growth and Development, University of Auckland, Private Bag 92019, Auckland, New Zealand, [3]AgResearch Limited, Animal Behaviour and Welfare, Ruakura Research Centre, Private Bag 3123, Hamilton, New Zealand

In sheep production on pasture, conception often occurs when ewes have poor nutrient reserves post-weaning and food availability is low. Periconceptional undernutrition (PU) alters endocrine and metabolic responses in the offspring. The aim of this study was to assess effects of ewe PU (10-15% body weight reduction) on offspring behaviour. We studied four nutritional treatments: well nourished (25 singletons, 14 twins) or undernutrition from -60 to +30 d from mating (18 singletons, 18 twins), -2 to +30 d (15 singletons) or -60 to 0 d (26 singletons). We evaluated lamb responses to maternal separation and preference for the dam over an alien ewe at 24h, 1 and 4 weeks of age using a triangular arena comprising two adjacent pens for ewes opposite a lamb pen. Fifty centimetres in front of both ewes' pens were designated as contact-zones (CZ). Ewe and lamb were placed in their pens and lamb vocalisations recorded (5min). An alien ewe was then placed in the second pen and the lamb's gate opened. Times to leave pen, reach CZ and spent in either CZ were recorded. Data were compared using repeated measures analysis allowing for nutritional group, twin/singleton, sex and year plus ewe as random effect. Female lambs vocalised more than males (62.8 vs 53.0, SED=4.7, p≤0.01) and left the pen faster (4.3 vs 6.2 s, SED=1.9, p≤0.01). Twins vocalised more than singletons (63.4 vs 52.5, SED=5.5, p≤0.05). Percentage of time spent in mother's CZ was similar in all groups. There were no significant differences amongst nutritional groups. These data suggest that female lambs show stronger attachment to their dam than males as early as 24h after birth and persisting to 4 weeks of age. Twin lambs may vocalise more than singletons due to competition for their mother's attention. PU had no effect on the parameters evaluated.

The relationship between thermal nociceptive threshold in lambs and mother-young interactions

S. Hild[1], I.L. Andersen[2] and A.J. Zanella[1], [1]Norwegian School of Veterinary Science, Department of Production Animal Clinical Sciences, PO Box 8146 Dep, NO-0033 Oslo, Norway, [2]Norwegian University of Life Sciences, Department of Animal and Aquacultural Sciences, PO Box 5003, 1432 Ås, Norway

Poor maternal care is likely to contribute to neonatal mortality in lambs. We hypothesised that, as primary caregivers, ewes can modulate their offspring's sensitivity to pain. We tested the sensitivity of lambs to thermal stimulus and studied the relationship between nociceptive thresholds and indicators of mother-young interactions. Our first goal was to validate an experimental tool to test thermal nociceptive thresholds in lambs. Our second goal was to study mother-young interactions using measures of proximity and synchronicity between individuals. Eight lambs were subjected to thermal stimulation at 10 or 11 days old. A telemetric device was used to gradually increase the temperature of a heating element placed on the hairless underside of the tail until a behavioural response, such as tail waging or lifting, was observed. A safety cut-off was set at 50 °C to prevent skin damage. Thermal stimulation was repeated every 30 minutes for three hours. Scan samplings were conducted every 10 minutes during the experiment. The distance between the ewe and lamb was measured as well as their synchronicity (performing the same activity). Threshold measurements yielded no significant individual differences among the tested lambs (PROC MIXED, SAS). The mean (±SD) differential between basal skin temperature and temperature at pain response was 4.5±2.2 °C. Overall thermal thresholds had no measurable effects on the behaviour of ewes and lambs. Nevertheless, the percentage of observations when the ewe and lamb stayed between 0 and 10 cm from each other was positively correlated with the highest nociceptive threshold (Pearson correlation: R=0.79, P<0.05). Likewise, activity synchronicity was positively correlated with the highest nociceptive threshold in lambs (R=0.76, P<0.05). These results indicate that increased physical proximity and increased behavioural synchronicity, which could be interpreted as indicators of positive mother-young interactions, are associated with a decreased sensitivity to thermal nociceptive stimulus in lambs.

The relationship between pre-pubertal inactivity and maternal ability in loose-housed sows

K. Thodberg[1], E. Jørgensen[2], A.C. Olsson[3], B. Houbak[1] and L.J. Pedersen[1], [1]University of Aarhus, Faculty of Agricultural Sciences, Department of Animal Health, Welfare and Nutrition, Blichers Allé 20 - P.O. Box 50, DK-8830 Tjele, Denmark, [2]University of Aarhus, Faculty of Agricultural Sciences, Department of Genetics and Biotechnology, Blichers Allé 20 - P.O. Box 50, DK-8830 Tjele, Denmark, [3]Swedish University of Agricultural Sciences, Department of Rural Buildings and Animal Husbandry, Box 59, S-230 53 Alnarp, Sweden

Most crushed piglets die within the first 24 h post-partum, emphasising the importance of the sow's behaviour. A connection between pre-pubertal and maternal behaviour has previously been found, showing that gilts, reacting by immobility in stressful test situations, spend more time in lateral recumbency and make fewer postural changes peri-parturiently. This study aims at testing the possibility of selecting sows with optimal maternal behaviour for a loose-housing system based on a 3-min novelty test at 3½ months of age. A total of 1529 gilts from three herds were tested. The 10 percent most active and least active individuals were selected, mated and moved to a loose-housing farrowing system one week before farrowing. Data are based on a total of 196 farrowings, of which 157 were video taped. The behaviour from 4 hours before the birth of the first piglet (BFP) until 12 h after BFP and piglet mortality was registered. Data were analysed using a graphical model which describes the conditional dependencies in the multivariate data structure. To avoid problems of multiple significance testing of correlations, an overall measure of model fit, the Bayesian Information Criteria (BIC), was used for model selection. Behaviour in the novelty test and behaviour around parturition were related through a weak positive relationship (partial correlation=0.19) between time spent walking without exploring and time spent in lateral recumbency before BFP. Thus only indirect dependencies were found between duration of inactivity in the novelty test and peri-parturient behaviour. The number of postural changes during farrowing and piglet mortality were positively related (partial correlation=0.25), but because variables recorded peri-parturiently were highly correlated, the relation might be via other variables, such as time spent in lateral recumbency. In conclusion, activity during novelty testing was not sufficient to predict piglet mortality, nor maternal behaviour. Thus, selection of sows with optimal maternal abilities, using this pre-pubertal behaviour, was not an efficient method.

Litter size and maternal investment in sows

I.L. Andersen[1], E. Nævdal[2] and P. Jensen[3], [1]Norwegian University of Life Sciences, Department of Animal and Aquacultural Sciences, P.O.Box 5003, 1432 Ås, Norway, [2]University of Oslo, Department of Economics, P.O.Box 1095 Blindern, N-0317 Oslo, Norway, [3]Linköping University, Department of Physics, Chemistry and Biology, 58183 Linköping, SE-58183, Sweden

The aim of the present study was to examine how litter size affected maternal behaviour and piglet survival and success in competition for teats. Forty healthy, Landrace x Yorkshire sows of different parity (1 to 4) and with a litter size ranging from 9 to 16 piglets were used. The piglets were weighed on day 1, 2.5 weeks and 5 weeks after birth (at the time of weaning), and all the dead piglets were subjected to a post mortem examination to ascertain the causes of death. Direct, behavioural observations of the sows and their litters were made on day 1, 2.5 weeks and 5 weeks after farrowing. General activity pattern of the sows were scored using instantaneous sampling with 5 minutes intervals during a 6-h observation period. The quality of 6 nursings was evaluated. The behaviour was analysed using a GLM-procedure in SAS with sow parity as aclass variable and litter size at birth as a regression variable. Increased litter size resulted in increased piglet mortality from birth until weaning ($P<0.001$), both due to crushing ($P<0.05$) and starvation ($P<0.05$), lower piglet weight at day one ($P<0.001$) and 5 weeks ($P<0.05$) after birth, and lower weight gain throughout the lactation period ($P<0.01$). Nursing frequency tended to increase with increasing litter size on day 1 after farrowing ($P=0.06$). Number of piglets not getting access to a teat ($P<0.01$) and not being present at the udder ($P<0.01$) during milk let-down both increased with increasing litter size on day 1 after farrowing. Sows giving birth to large litters spent more time on activities not involving the piglets, such as standing ($P<0.05$), moving ($P<0.05$), rooting on the floor/litter ($P<0.05$) and were generally more active with the piglets in close proximity. In conclusion, an increased litter size resulted in more piglets being crushed and starved, an increased number of piglets not getting access to a teat, and that the sows spent less time on piglet-directed activities.

Space use after introduction of a single or a pair of heifers into free stall dairy herds

L. Gygax[1], G. Neisen[1,2] and B. Wechsler[1], [1]Federal Veterinary Office, Centre for proper housing of ruminants and pigs, Agroscope Reckenholz-Tänikon Research Station ART, Tänikon, 8356 Ettenhausen, Switzerland, [2]University of Münster, Department of Behavioural Biology, Badestr. 13, 48149 Münster, Germany

The introduction of heifers into herds of dairy cows is a frequent but sparsely investigated event. We introduced a single (singleton) and a pair of heifers (pair) in a balanced order on six farms. Using an automatic tracking system, we continuously recorded positions of the introduced heifers as well as all cows once per minute over the first 6 days after introduction. For each individual and day, we calculated the proportion of time spent in the activity, the feeding and the lying area, the path length and the size of the area used within the free stall system. We compared the measures of space use (1) of heifers and cows in the introductory weeks and (2) data of the cows from the control weeks preceding the introduction with the introductory weeks using linear mixed-effects models. Singletons spent more time in the activity ($F_{2,182}=23.3$, $p<0.001$) and less time in the lying area compared to pairs and cows (typexday: $F_{6,1319}=2.8$, $p=0.011$). Average path length was increased for the introduced heifers early in the week but approached the values of the cows towards the middle (pairs) and end of the week (singletons, typexday $F_{6,1317}=2.7$, $p=0.012$). The area used was largest in the cows and was clearly reduced in the singletons and slightly less so in the introduced pairs ($F_{2,182}=3.0$, $p=0.05$). The space use of the cows comparing the control and introductory weeks were influenced to a greater extent by the pairs than by singletons (typexday: all $F_{6,2477-2484}>4.0$, $p<0.01$) but the differences were small on an absolute scale. In conclusion, heifers introduced as a pair had a space use more similar to cows of the herd than singly introduced heifers and effects on space use of the cows by introduced heifers was minor. We would thus recommend to integrate pairs rather than single heifers.

Grouping primiparous dairy cows: behaviour, production performance and cortisol in serum and hair

M. González-De-La-Vara[1], F. De-Anda[2], J.C. Vazquez-Ch[3], A. Romero[4], M.C. Romano[5] and A. Villa-Godoy[6], [1]UNAM, Ethology and Wild life. Faculty of Veterinary Medicine, Circuito exterior SN, 04510 Mexico City, Mexico, [2]Agropecuaria El Gigante, Management, Km 25 Carr. Aguascalientes- Zacatecas, 20303, San Francisco de los Romo, Mexico, [3]UAEM, CIESA, Autopista Toluca-Atlacomulco Km 15.5, 50200, Toluca Edo. Mex., Mexico, [4]Agropecuaria El Gigante, Reproducción, Km 25 Carr. Aguascalientes- Zacatecas, 20303, San Francisco de los Romo, Mexico, [5]CINVESTAV, Fisiología, Biofísica y Neurociencias, IPN 2508, 07360, Mexico City, Mexico, [6]UNAM, Fisiología y Farmacología, College of Veterinary Medicine, Circuito Exterior S/N, 04510, Mexico City, Mexico

In order to reduce social stress and improve productivity, it has been suggested to group primiparous with primiparous cows at the beginning of the lactation. The aim was to compare behaviour, productivity and cortisol levels in hair and serum of primiparous grouped with or without multiparous cows. 44 Holstein cows were randomly assigned in two groups: PP, 22 primiparous with 128 primiparous cows; PM, 22 primiparous with 128 multiparous cows. Animals were housed with maximal comfort and avoiding stress as much as possible. Integral diet and water were provided on an ad libitum basis. Social, sexual and maintenance behaviour were observed directly for 4 hours/day from the 1st to 90th day of lactation. On days 0, 30, 60 and 90 of lactation, body condition and somatic cell count (SCC) were measured and blood and hair samples were taken to analyse cortisol by RIA. At milking, body weight, milk yield, activity (podometer) and milk conductibility was recorder from day 0 to 180 of lactation. Behavioural data were analysed by Mann-Whitney U test and production data with ANOVA. PP show higher frequencies of aggression performed (PP:3.32, PM:1.73, P=0.01) and displacement (PP:4.05, PM:1.59, P=0.0003) than PM. No differences were found (P>0.05) between groups in activity, number of times observed standing (PP:10.91, PM:9.05), walking (PP:0.91, PM:1.18), eating (PP:18.5, PM:15.36), ruminating (PP:6.82, PM:6.68), lying (PP:23.05, PM:24.14) or drinking (PP:0.82, PM:0.55). No differences were found between groups (P>0.05) in milk yield (PP:31.8, PM:32.1Kg/day); days from birth to conception (open days) (PP:115.8, PM:102.5), SCC, milk conductibility and hair cortisol (PP:7.94, PM:7.2ng/mL). PP serum cortisol was higher at 60 and 90 day of lactation (PP:3.34, 8.56; PM:6.75, 5.57ng/mL, P<0.03). Body condition was higher in PM at day 90 of lactation than PP (PP:2.58, PM:2.96, P=0.04). Thus, in these conditions of resources, management and space allowance, grouping primiparous with multiparous cows is recommended.

Effects of ractopamine feeding, gender and social rank on aggressiveness and monoamine concentrations in different brain areas of finishing pigs

R. Poletto[1,2], H.W. Cheng[2], R.L. Meisel[3], B.T. Richert[1] and J.N. Marchant-Forde[2], [1]Purdue University, Animal Sciences, West Lafayette, IN 47907, USA, [2]USDA-ARS, Livestock Behavior and Research Unit, West Lafayette, IN 47907, USA, [3]Purdue University, Psychological Sciences, West Lafayette, IN 47907, USA

Aggression in pigs is of great concern due to its negative impact on production and welfare. Our goal was to evaluate the effects of the feed additive ractopamine (RAC), gender and social rank on aggressiveness and levels of brain monoamines in finishing pigs. A total of 16 barrows and 16 gilts were assigned as either control (CTL) or RAC feeding at 5ppm for 2wk, then 10ppm for another 2wk (pen=4/gender). Dominant and subordinate pigs in each pen were determined over 36h post-mixing. Aggressiveness was measured by latency to first bite and cumulative frequency of attacks over 300s using 6 resident-intruder tests performed on the wk prior to start, wk 2 and 4 of the trial. At wk 4, frontal cortex (FC), hypothalamus (HYP), amygdala (AMY) and raphe nuclei (RN) were analysed for serotonin (5HT), dopamine (DA), their metabolites, norepinephrine (NE) and epinephrine (EP) levels using HPLC. A 2×2×2 analysis with repeated mixed models was computed and adjusted by Tukey post-hoc test. Latency to attack tended to be shorter in RAC-fed than CTL pigs (83.6 vs. 115.5±16.6s; $F_{1,22}$=3.03 p=0.09). By 60s, dominant and subordinate RAC-fed gilts had performed more attacks (54.17% and 66.67% respectively) compared to other individuals ($F_{1,12}$=20.7 p<0.001), and CTL dominant gilts had performed greatest percentage of attacks (96.7%) by 299s ($F_{1,12}$=16.7 p<0.01). Serotonin levels were lower in the FC of gilts ($F_{1,24}$=4.7 p<0.05) and in RN of dominant ones ($F_{1,24}$=5.1 p<0.05). When fed RAC, gilts also showed lowest 5HT metabolite levels ($F_{1,24}$=4.0 p<0.05) and greatest DA turnover in AMY ($F_{1,24}$=7.7 p<0.05). Dominant barrows had higher EP than dominant gilts in HYP ($F_{1,24}$=4.5 p<0.05). Serotonin deficiency and enhanced DA metabolism in brain areas essential for aggression inhibition may be associated with greater impulsive aggression observed in gilts, especially when fed RAC. Although RAC is a valuable feed additive for enhancing growth performance in swine, its welfare effects must be considered.

A comparison of regrouping strategies affecting aggression in recently grouped sows

S.M. Hayne and H.W. Gonyou, Prairie Swine Centre, 21057-2105-8th Street East, Saskatoon, SK, S7H 5N9, Canada

A study was conducted using 18 groups of 16 sows each to study the aggression and injuries resulting from group formation in an 80m^2 pen after breeding. The treatments (3-4 groups/treatment) included 'Control', 'Familiar' (previously housed together), 'Dominant' (regrouped with three dominant sows to the study animals), 'Exposed' (mixed for 48h prior to being stalled for breeding), and 'Protected' (access to protective stalls). Data were analysed using Proc Mixed in SAS with group as the experimental unit. Overall, the number and duration of fights and injuries did not differ among the treatments. Salivary cortisol concentration also did not differ among the treatments. More fights tended to occur on the first day of regrouping compared to the following two days (Day 1: 10.2, Day 2: 4.7, Day 3: 7.5; $P<0.1$). The duration of fights also decreased over time, with the longest fights occurring on the first day (Day 1: 17.3s, Day 2: 5.9s, Day 3: 3.6s; $P<0.05$). At 24h post-regrouping, injuries were greater in the Control (5.0) and Dominant (4.9) compared to the Exposed (3.4) ($P<0.05$), and tended to be greater than in the Familiar (3.8) treatment ($P<0.1$). At 48 and 72h post-regrouping, injuries were greater in the Control (5.3) compared to Dominant (3.9), Exposed (3.3), Familiar (3.3) and Protected (3.9) ($P<0.05$). After 10d, injuries were greatest in the Control (3.1) and Dominant (3.2) compared to the Exposed (2.0), Familiar (2.2) and Protected (1.9) treatments ($P<0.05$). Cortisol increased following regrouping, and remained higher than pre-regrouping values (Pre:3.9nmol/L, Post-24h:5.9nmol/L, Post-48h:6.6nmol/L, Post-72h:5.6nmol/L, Post-10d:6.5nmol/L; $P<0.05$). Although there were no overall treatment effects on the measured variables, differences in injuries appeared over time. This could indicate that more subtle aggression and chronic stress persisted in the Control and Dominant treatments, and social stability occurred faster in the Familiar, Exposed and Protected treatments.

Visual recognition of familiar conspecifics; can heifers recognise herd members?

M. Coulon[1], B.L. Deputte[1,2], Y. Heyman[3] and C. Baudoin[1], [1]Laboratoire d Ethologie Expérimentale et Comparée, Université Paris 13, 99 avenue JB Clément, F-93430 Villetaneuse, France, [2]ENVA, Ethology, 7 avenue Général de Gaulle, F-94704 Maisons-Alfort Cedex, France, [3]INRA, Biologie du Développement et Reproduction, domaine de Vilvert, F-78352 Jouy-en-Josas, France

Cattle are often regrouped causing repeated changes in social organisation. Re-establishing group stability may be achieved by rapid individualisation of group members by each individual. Rapid knowledge and recognition of group members may be a key factor of social adaptability. This study aimed at evaluating the capacity of individual recognition in Prim' Holstein heifers, using only visual cues. Subjects were confronted with 2D images of either familiar (experiment-1) or unfamiliar heifers (experiment-2). Stimuli included natural-size heads of cows under different orientations (face, profile, etc.). Experiments were based on simultaneous S+/S- discrimination through instrumental conditioning with food rewards. Half the subjects were initially assigned to experiment 1 and the remainder to experiment-2. In the training phase of both experiments, the same pair of stimuli was used in each trial, the face of the "target "cow and the face of another "distractor" individual. In the generalisation phase, subjects had to recognise new images of the target individual (different orientations). The distractor stimuli comprised various images of three cows. The criterion was set at least 80% of good responses per session of 10 trials, during 2 consecutive sessions. The results showed that all heifers reached the criterion in 2 to 10 sessions. They learned to recognise images of a familiar individual significantly faster than those of an unfamiliar one (generalisation test: 3.12±0.66 vs. 5.25±1.17 sessions; Wilcoxon test, N=8, z=1.99, p=0.046, two-tailed). In our study, all the heifers treated the different views of one individual as if they were equivalent to each others, indicating a capacity of individual recognition. A social experience with direct interactions between heifers facilitated the recognition process. Heifers showed also their ability to recognise images of unfamiliar individuals, revealing their ability to quickly adapt to new social environments.

Allogrooming in cattle: influence of social preferences, social dominance and feeding competition

D. Val-Laillet[1], V. Guesdon[1], M.A.G. Von Keyserkingk[2], A.M. De Passillé[1] and J. Rushen[1], [1]Pacific Agri-Food Research Centre, Ethology Laboratory, 6947 #7 Highway P.O. Box 1000, V0M 1A0 Agassiz, BC, Canada, [2]University of British Columbia, Animal Welfare Program, 2357 Main Mall, V6T 1Z4 Vancouver, Canada

Little is known of the social functions of allogrooming in cattle. To describe the relationship between allogrooming, social dominance and feeding competition, we observed the occurrence of allogrooming, displacements at the feed bunk and spatial associations between animal dyads in 6 groups of 8 dairy cows continuously for 4d. Then, the level of social competition was increased by halving the access to the feed bunk during 4d. Allogrooming was directed mainly to the head (35%) and neck (45%) and occurred mostly at the feed bunk and in the nearby alley (74%). Dominance rank had no effect on the occurrence of allogrooming ($P>0.10$). Each cow groomed 4.5 ± 0.3 herd members out of 7. 18% of the dyads showed no allogrooming while 45% of dyads had reciprocal grooming (the last 37% were one-way dyads). Dyads composed of two multiparous cows spent more time allogrooming than mixed dyads ($P<0.02$). Most frequent neighbours at the feed bunk were found to groom each other more (98 ± 37 sec at the feed-bunk and 125 ± 35 in the alleys) compared to less frequent neighbours (reciprocally 39 ± 11 sec and 50 ± 19 sec) ($P<0.001$). When competition between animals was increased, the frequency of displacements at the feeder increased (12 ± 1 vs. 22 ± 2 displacements per animal, $p<0.001$) and the daily duration of time spent allogrooming decreased (320 ± 55 vs. 193 ± 28 s per animal, $P<0.05$), especially in low-ranking, primiparous animals ($P<0.01$). Allogrooming may be a behaviour reflecting social affiliation in cows and is influenced by social dominance. Primiparous cows may be more susceptible to suffer from a lack of socio-positive relationships when submitted to high competitive pressure, especially when they are mixed with more experienced animals.

Discrimination between conspecific odour samples in the horse (*Equus caballus*)

B. Hothersall[1], P. Harris[2], L. Sortoft[3] and C. Nicol[1], [1]University of Bristol, Division of Clinical Veterinary Science, Langford, North Somerset, BS40 5DU, United Kingdom, [2]WALTHAM Centre for Pet Nutrition, Equine Studies Group, Melton Mowbray, Leicestershire, LE14 4RT, United Kingdom, [3]Hartpury College, Hartpury, Gloucestershire, GL19 3BE, United Kingdom

Discrimination between odour cues from different conspecifics has been demonstrated in various rodents, pigs and some primates. In horses, behavioural observations suggest that smell is important in social interactions but balanced studies of discriminative capacity are lacking. We used a habituation-discrimination procedure to investigate the ability of pregnant mares, of mixed breed, to distinguish between pairs of odour samples from different individuals. Pairs to be discriminated always consisted of one young gelding and one pregnant mare. Separate tests were conducted for urine, faeces and secretions from the neck and face rubbed onto fleece fabric. 10 mares each underwent three separate tests, one per sample type. A test consisted of three successive two-minute presentations of a small sieve containing a sample from Individual A, with inter-trial intervals of 15 minutes. During the final presentation only, a separate sample of the same type from Individual B was simultaneously presented. The duration of all investigative responses to the sample/s was recorded. Doubly repeated measures ANOVA indicated that sample type affected duration of investigative responses towards the repeated sample (F_2=7.797, p=0.004): durations were longer for secretions than faeces or urine. Response durations declined across the 3 presentations (F_2=18.083, p<0.001). Donor sex and side of presentation were balanced across subjects and tests; further ANOVAs on each sample type showed both were non-significant. We examined whether mares investigated the novel sample (B) more than the repeated sample on the final presentation – thus demonstrating discrimination – using Wilcoxon's matched pairs tests, as variance was unequal for novel and repeated samples. Mares investigated the novel sample more for urine samples only (novel: median=2.5s, IQR=6; repeated: median=8.0s, IQR=10; Z=-2.558, p=0.008). These findings suggest that horses are able to discriminate between conspecifics by urine odour, at least at the level of broad categories such as sex or reproductive status.

Are long lasting social relationships significant for calves?

S. Raussi[1], S. Niskanen[2], I. Veissier[3], J. Siivonen[4], L. Hänninen[5], H. Hepola[6] and L. Jauhiainen[7], [1]MTT Agrifood Research Finland, Animal Production Research, Vakolantie 55, 03400 Vihti, Finland, [2]University of Helsinki, Department of Ecological and Environmental Sciences, Viikinkaari 9, 00014 University of Helsinki, Finland, [3]INRA, INRA-Clermont-Ferrand-Theix, 63122 Saint-Genès-Champanelle, France, [4]MTT Agrifood Research Finland, Animal Production Research, 31600 Jokioinen, Finland, [5]University of Helsinki, Department of Production Animal Medicine, Koetilantie 7, 00014 Univesity of Helsinki, Finland, [6]University of Helsinki, Department of Animal Science, Koetilantie 5, 00014 University of Helsinki, Finland, [7]MTT Agrifood Research Finland, Services Unit, 31600 Jokioinen, Finland

Calves develop long-lasting social relationships with peers. We examined the significance of the length of the relationship and its effects on calf behaviour in stressful situations. Twenty-four female dairy calves were reared in single pens for two weeks then assigned to six groups of four animals of the same age. The four animals were reared together until 1.5 years of age (Type-1 partners). At 14 weeks of age they were mixed with other calves of the same age (Type-2 partners) to form a group of 15 calves. Type-3 partners were calves added to the experimental group after 21 weeks. The position and activity of animals in the barn and pasture were scanned every 10 minutes for 6 hours on two consecutive days and in four periods. Behavioural synchrony, distance between animals and proximity were analysed. Social preferences of calves between three partner types were examined in a Y-maze and the calming effect of Type-1 and unfamiliar partner in an arena test. Data were analysed with SAS mixed model and Kruskal-Wallis-tests. A subject calf spent more time close to the Type-1 partners than Type-2 or Type-3 partners (15.2% vs. 12.5% or 7.9% of time, F=16, p<0.001). Calves butted Type-3 or unfamiliar partners more frequently than Type-1 partners in the Y-maze and in the arena (1 vs. 0 and 9 vs. 3 times, $\chi^2 \geq 6$, p<0.05). Calves vocalised more when in the arena alone than with Type-1 or unfamiliar partner (25 vs. 8 or 9, F=33, p<0.001). Calves walked less in the arena when they were with Type-1 partner than with an unfamiliar partner or alone (125 vs. 177 or 151 s, F=7, p<0.01). Calves seem to form strong relationships before 14 weeks of age. These relationships reduce further aggressions and responses to novelty, and may thus help calves to cope with stressful situations.

Physiological assessment of emotional reactions in sheep

N. Reefmann[1,2], F. Bütikofer[1,3], B. Wechsler[1] and L. Gygax[1], [1]Federal Veterinary Office, Centre for Proper Housing of Ruminants and Pigs, ART Tänikon, 8356 Ettenhausen, Switzerland, [2]University of Münster, Behavioural Biology, Badestr. 13, 48149 Münster, Germany, [3]ETH Zürich, Institute of Animal Sciences, Physiology and Behaviour, Universitätsstr. 2, 8092 Zürich, Switzerland

With the aim to judge emotional valence from an animal's perspective, this study recorded multiple physiological parameters in sheep exposed to situations likely to induce negative and positive emotional states. Fourteen sheep were trained to anticipate the delivery of standard feed. After several weeks of training, each sheep was exposed to positive-negative-contrasts: animals were either offered familiar standard feed (control treatment), unpalatable wooden pellets (negative treatment), or energetically enriched feed (positive treatment). Heart rate, root-mean-square-successive-difference (RMSSD), respiratory frequency, skin temperature, skin humidity and eye white were recorded during 6 minutes prior to the delivery of feed (anticipation) and for 6 min during the delivery of either standard feed, wooden pellets, or enriched feed. Data were analysed using linear mixed-effect models. Heart rate ($F_{2,35}=5.39$; p=0.009), respiratory frequency ($F_{2,33}=5.82$; p=0.007) and variability of skin humidity ($F_{2,39}=16.37$; p<0.001) showed a similar pattern: values were high during the delivery of wooden pellets and low during the delivery of standard or enriched feed, whilst an inverse pattern was observed for RMSSD ($F_{2,35}=2.40$; p=0.11). Of these measures, heart rate was additionally influenced by the sequence of testing of the different treatments, with a higher heart rate during the first exposure to either wooden pellets or enriched feed ($F_{2,23}=9.96$; p<0.001). Skin temperature ($F_{2,39}=0.64$; p=0.53) and eye white ($F_{2,3}=0.94$; p=0.48) were not influenced by treatment. These results indicate that negative emotional states coincide with cardio-respiratory acceleration and an increase in variability of perspiration. Both the control and the positive treatment elicited cardio-respiratory deceleration and reduced perspiration-variability. Sheep may have perceived these treatments as positive, given that their expectations of feed were fulfilled in both with the delivery of feed. Thus, the negative treatment was clearly differentiable from the other two treatments, and heart rate, respiration and skin humidity may help to judge emotional valence from a sheep's perspective.

Heart rate and heart-rate variability in domestic goats whilst feeding and during social separation

J. Aschwanden, L. Gygax, B. Wechsler and N.M. Keil, Centre for proper housing of ruminants and pigs, Federal Veterinary Office, Tänikon, 8356 Ettenhausen, Switzerland

Heart-rate variability is increasingly used to assess stress in farm animals. In this study we investigated whether cardiac activity in terms of heart rate (HR) and root mean square of successive beat-to-beat differences (RMSSD) is influenced in domestic goats by rank, feeding distance and social separation. During feeding side-by-side, HR and RMSSD were non-invasively recorded with a three-channel digital Holter-Lifecard-CF in five dyads each of eight different goat groups with known rank relationships. Each dyad was observed while feeding in a "far" (the dyad's freely chosen distance) and in a "near" (the dyad's minimum distance not eliciting agonistic interactions) feeding distance. Baseline values measured before the feeding tests as well as the differences between the baseline and test values (D) in HR and RMSSD were analysed with linear mixed-effects models with crossed random effects. Furthermore, D-values of HR and RMSSD obtained by measuring baseline prior to social-separation tests and test values during social separation carried out with the same individuals were compared to the D-values of HR and RMSSD obtained in the feeding tests. Goats with a high rank within the group had higher levels of baseline-RMSSD and lower levels of HR than goats with a low rank within the group (p<0.05). With regard to D-RMSSD, we found a significant interaction between feeding distance and rank within the group (p=0.01). Low-ranking goats had lower D-values at the far than at the near distance, whereas the converse was true for high-ranking goats. In the separation tests, D-HR was significantly higher (55.7±4.8 vs. 18.2±1.3, p<0.05), whereas D-RMSSD did not differ from the D-values observed in the feeding tests (7.4±5.0 vs. 14.1±4.0). In conclusion, our results demonstrate that cardiac response in goats is not only influenced by the context (feeding vs. separation) but also by the individual's rank within the group.

Stress coping in farmed Atlantic salmon (*Salmo salar*)

S. Kittilsen, Ø. Øverli and B.O. Braastad, Norwegian University of Life Sciences, Department of Animal and Aquacultural Sciences, P.O. Box 5003, 1432 Aas, Norway

Farmed fish are exposed to an environment that may be unpredictable and uncontrollable with several potential stressors. The ability to cope with stress may therefore contribute to both better welfare and growth. The aim of this study was to investigate differences in stress coping between family groups of Atlantic salmon. Behavioural and physiological responses to confinement stress were monitored on an individual level in 91 juvenile salmon from 10 families (n=10). Locomotor activity and water-borne cortisol secretion during acute stress were analysed. Mean family values were related to central parameters from the breeding company's registers such as selection value for growth and disease resistance. Aggressive behaviour when faced with a social stress situation was also investigated. Average family values for water-borne cortisol concentrations correlated positively with locomotor behaviour during confinement (R^2=0.72, p=0.002). This supports the idea that more stress sensitive fish move more in an acute stress test. Our results also revealed a correlation between locomotor behaviour during confinement and disease resistance towards infectious pancreatic necrosis virus (IPN) (R^2=0.58, p=0.01). Families that react with high amounts of locomotor activity to confinement show higher mortality in disease challenge tests. Further a non-linear regression was detected between aggression and the corresponding mid parent selection-value for growth (R^2=0.62). Data were found to be non-linear by Runs test. Best fit were found by using 2.order polynom (Microsoft Exel). Offspring of both the most slow-growing and the fastest growing parents are characterised by enhanced aggression towards an intruder. It appears that non-invasive behavioural tests performed in the laboratory have predictive value regarding how the same families will perform in the full scale breeding program. Selection could create the basis for more stress tolerant individuals. This might be beneficial for the aquaculture industry both when it comes to production parameters like growth, and individual welfare criteria.

Transmission of stress-induced adaptive behaviour from parents to offspring in chickens

D. Nätt and P. Jensen, IFM Biology, Division of Zoology, Linköping University, 58183 Linköping, Sweden

We have previously shown that offspring of stressed domestic chickens can inherit learning impairments of their parents and the associated modifications in brain gene expression. The aim of this study was to study possible adaptive aspects of such cross-generation transmission. We hypothesised that stress would cause chickens to prefer predictable food and to be more dominant, which would be transmitted to the offspring. Commercial Hyline birds (n=15-16/treatment) were raised in an unpredictable light rhythm (stress treatment) or in control conditions (12:12 h light:dark). Stressed animals pecked more at a safe food source than at a hidden source compared to birds from the control group (mean ± SEM: 26.1±4.7 vs 11.1±3.5 pecks/min). Female (not male) offspring (n=16) of stressed birds, raised in control conditions without parental contact, also pecked more at the safe source compared to offspring (n=20) of control parents (14.9±2.5 vs 9.1±1.4 pecks/min). Furthermore, adult offspring of stressed birds (n=10-22/treatment) performed more food pecks in a competition test (7.5±2.3 vs 1.8±4.4 pecks/min), showed a higher preference for high energy food over standard food (51.4±4.2 vs 28.9±6.8% of time occupancy) and were heavier at day 154 (1682±52 g vs 1551±43 g) than offspring of control parents. Differences were analysed with GLM or paired-sample T-test, and all significant results reported had at least $P<0.05$. Using cDNA microarrays, brain gene expression differences between treatments were analysed in each generation. There was no global correlation between expression profiles in parents and offspring, but several genes were found on the top lists in both generations and were also identified from our previous experiment. Our findings suggest that stress may cause adaptive responses in feeding behaviour, which may be transmitted to the offspring by means of epigenetic mechanisms. This may in turn prepare the offspring for coping with an unpredictable environment.

Proportion of time spent lying, standing and active by grazing beef heifers differing in phenotypic residual feed energy intake

P. Lawrence[1,2], M. McGee[1], M.J. Drennan[1], D.A. Kenny[2], D.H. Crews Jr.[3] and B. Earley[1], [1]Teagasc, Grange Beef Research Centre, Dunsany, Co. Meath, Ireland, [2]University College Dublin, School of Agriculture, Food Science and Veterinary Medicine, Belfield, Dublin 4, Ireland, [3]Agriculture and Agri-Food Canada Research Centre, Lethbridge, Alberta, Canada

Cattle differing in feed efficiency may exhibit different behavioural activity. The effect of residual feed intake (RFI), an alternative measure of feed efficiency, on the proportion of time spent lying, standing and active at pasture was determined using 63 Simmental and 21 Simmental x Friesian-Holstein pregnant heifers. As weanlings during the winter indoor period, live weight and individual feed consumption records were used to generate RFI values (expressed as Unité Fourragère Lait (UFL) - feed units for lactation) within each genotype. The heifers were ranked on RFI and divided equally into either low (-0.32 UFL, efficient) or high (+0.27 UFL, inefficient). At the end of the indoor winter period, they were turned out to pasture (April) where the low and high RFI groups were rotationally grazed separately as 4 (2 replications) herds of 21 animals on predominantly perennial ryegrass swards. Mean paddock size was 1.02 (0.56 to 1.42) ha. The heifers were bred to Simmental sires. Time spent lying, standing and active was determined over 3 consecutive days (following one day adaptation) during September (mean live weight 515 kg) using pedometers (IceTag 2.004, IceRobotics Ltd.) positioned on the left back leg. Data were analysed using GLM, with a model that included RFI group, grazing herd and genotype. Birth day was included as a covariate. The percentage of time spent lying, standing and active, and the number of steps taken daily (s.e.m.) were 47.8 and 48.0 (1.04), 46.4 and 46.2 (1.04), 5.8 and 5.8 (0.23), and 3926 and 3855 (203.8) for low and high RFI groups, respectively. Behaviour at pasture was similar (P>0.05) between the RFI groups. In conclusion, the absence of a difference between the RFI groups indicated that energy requirements associated with standing and walking did not contribute to the differences in feed efficiency.

The reduction in the frequency of beef cattle supplementation of feed: behaviour and performance

M.J.R. Paranhos Da Costa, M.H. Quintiliano and A.G. Páscoa, Faculdade de Ciências Agrárias e Veterinárias, UNESP, Departamento de Zootecnia, Via de Acesso Dr. Paulo Donato Castellane Km 5, 14884-900, Brazil

The aim of this study was to test the hypothesis that the reduction in the frequency of feed supplementation would affect the social behaviour and weight gain in Nelore steers. The study was carried out with 48 Nelore steers, divided in six groups. All groups were maintained in pastures and received the same feed supplementation. Three of these groups received the supplementation daily (DS=1Kg/animal/day) and for the others the supplement was offered three times per week (RS=2kg/animal every Monday and Wednesday and 3kg/animal every Friday). The animals were fed in these regimes for 180 days. They were weighed every 28 days, evaluating their weight gain (WG). The steers' behaviour was recorded weekly, for a twenty four hour period. Social interactions and supplement ingestive behaviour were recorded by direct and continuous observations, using focal sampling. Instantaneous sampling (15 min. interval) was adopted to record grazing, resting and rumination behaviour. For this approach, two behavioural variables were considered: the percent of time in front of the feed bunk (TFB) and the frequency of physical aggression (FPA), defining which animal attacked (FPA-A) and which was attacked (FPA-R). One-way Anova was used to test the effects of treatments on WG, TFP and FPA. Spearman Rank correlation coefficients were estimated considering all variables. The treatments did not affect WG (Anova, $F_{1, 47}$=0.01, P=0.949: DS=0.21±0.41 and RS=0.21±0.06 Kg/day) and TFP (Anova, $F_{1, 47}$=2.46, P=0.124: DS=0.25±0.11 and RS=0.20±0.14%). As expected FPA was higher in RS than in DS (Anova, $F_{1, 47}$=3.104, P=0.085: DS=5.40±2.9 and RS=7.5±5.1 fights/day). There were positive correlations between TFB and both FPA-A and FPA-R (P<0.01: r=0.524 and r=0.456, respectively). In conclusion, the reduction in the frequency of feed supplementation increased the frequency of agonistic interactions among the steers, but did not affect their performance.

Postpartal behaviour and claw lesions in dairy cows

S. Dippel[1], C.B. Tucker[2], C. Winckler[1] and D.M. Weary[3], [1]BOKU - University of Natural Resources and Applied Life Sciences Vienna, Department of Sustainable Agricultural Systems/Livestock Sciences, Gregor-Mendel-Str. 33, 1180 Vienna, Austria, [2]University of California, Animal Science Department, 1 Shields Ave., Davis, CA 95616, USA, [3]University of British Columbia, Animal Welfare Program, 2357 Main Mall, BC V6T 1Z4 Vancouver, Canada

Behaviour is thought to play a role in the development of claw lesions such as haemorrhages (HAEM) and white line separation (disintegration between sole and wall horn, WLS). In order to understand how behaviour after calving affects the development of HAEM and WLS in dairy cows, we observed 23 multiparous and 9 primiparous cows housed in 4 groups in identical cubicle pens with solid concrete flooring and sand bedding. Two groups were housed at higher stocking density (0.75 instead of 1.0 stalls/cow) in order to reduce lying time. Lying, feeding and standing behaviour was recorded on 2 to 3 weekends (48h) within 21 days after calving (10-min scan sampling from video). Time budgets were summarised as mean. All claws were scored at 4, 8 and 12 weeks post partum. For each time point and cow we calculated four hoof health scores by adding up the respective lesion severity scores: total HAEM (HAEM-T), HAEM sole only (HAEM-S), HAEM white line only (HAEM-W), and total WLS (WLS-T). We used linear mixed models with repeated observations and HAEM-T, HAEM-S, HAEM-W and WLS, respectively, as dependent variables (random=cow nested in group). Mean time budgets were included as fixed factors and parity nested in stocking density as confounder. HAEM-T decreased by 3.0 when average daily feeding time increased by 1 hour (p=0.047). HAEM-T and HAEM-W were significantly confounded by parity nested in stocking density (p<0.05). HAEM-S was only significantly influenced by standing half in cubicles (p=0.008). It increased by 1.9 per additional daily hour standing with front legs in cubicles. There were no significant effects for WLS-T. Thus, postpartal behaviour is a risk factor for the development of claw haemorrhages in dairy cows. Barns should be designed and managed to increase feeding time and minimise time spent standing half in cubicles.

Straw bedding maintenance: effects on lying behaviour and preferences of dairy cattle

K. Reiter, M. Abriel and F. Freiberger, Institute for Agricultural Engineering and Animal Husbandry, Bavarian Research Center for Agriculture, Prof. Dürwaechter Platz 2, 85586 Grub, Germany

In cubicle houses for dairy cattle were use straw bedding or lying mats. Often farmers haven't enough time for cleaning and littering the straw beddings. Consequently the lying areas are less comfortable. This study aimed to comparing different levels of straw bedding (level 1: straw bedding 20 cm; level 2: bedding 5 cm; level 3: bedding 5 cm, around 3600 cm² without bedding) to soft lying mats (Kraiburg KEW PLUS) regarding cow behaviour and preferences in three consecutive stages. Data were collected in a dairy house with 60 cows (Bavarian Fleckvieh) and a robotic milking system. The lying behaviour and preferences of animals were filmed 24 h for 5 days (each stage) by digital video recorders. The observations started after 21 days, when the cattle were accustomed to the new level of bedding. In addition standing up, lying down behaviour and positions of lying were quantified by direct observations during 15 days, two observations per day two hours each. The nonparametric Friedman-test and the Mann-Whitney-U-test were used to analyse the data. Significant differences were detected in preference of straw bedding level 1 and 2 to soft lying mats (Average cubicle usage (24 h) 36.1% vs. 18.6% and 32.1% vs. 17.9%) but in level 3 cows showed preferences for soft lying mats (29.6% vs. 15.5%). The lying time (h/24 h) decreased from level 1 to 3 (13.2; 12.2; 10.9; p=0.034). Mean duration of lying bouts (minutes) was significantly shorter only in straw bedding level 3 compared to soft lying mats (level 1: 72.6 vs. 88.5; level 2: 81.2 vs. 90.7, level 3: 56.6 vs. 90.4; p=0.029). The duration of both lying down and standing up movements did not differ significantly between straw bedding and lying mats, as well as between the levels of straw bedding. The percentage of cows in half side lying position was significant higher in straw bedding level 3 compared to level 1 and 2 (6.4%, 6.8%, 13.2%). In conclusion reduced bedding levels affects the cows lying behaviour and lying position. High straw bedding quality is important for a good welfare in dairy cattle. If farmers have not enough time for cleaning and littering the straw beddings they should use soft rubber mats.

The influence of alteration of keeping system on milking cows´behaviour patterns

A. Pavlenko[1], T. Kaart[2], V. Poikalainen[3], L. Lidfors[4] and A. Aland[1], [1]Veterinary Medicine and Animal Sciences, Animal Helath and Environment, Kreutzwaldi 62, 51014, Estonia, [2]Veterinary Medicine and Animal Sciences, Animal Genetics and Breeding, Kreutzwaldi 46, 51006, Estonia, [3]Veterinary Medicine and Animal Sciences, Food Science and Hygiene, Kreutzwaldi 58A, 51014, Estonia, [4]Faculty of Veterinary Medicine and Animal Science, Animal Environment and Health, P.O. Box 234, Skara, 532 23, Sweden

The aim of this study was to investigate how quickly dairy cows learned to use resources such as food, water and lying places when they were mowed from a tied system to a cubicle system. The study was carried out on a loose housing farm with 1200 cows of Estonian Holstein Breed. A newly built barn with cubicles and concrete alleys and a milking carousel with 40 places were used. There were eight groups of cows with 100-120 animals in each. The cows were milked twice daily and had ad lib. access to a total mixed ration and water. The study was conducted over five weeks with 4 four-day observation periods each group was observed three times per day from 8.30 to 14.00 between milking times. All the following behaviours were recorded: lying, eating, walking, sleeping, ruminating and vocalising. Also the location of cows within the group area, when they stood, ruminated or lay, was recorded. The social behaviours of aggression and allogrooming were also recorded. Preliminary results from the study showed that at the start of the study cows lay on the walking area significantly more than in the end of the study (3.02% and 0.81% respectively). Lying, eating, drinking and sleeping behaviours were significantly related to the time of the day; e.g. cows lay significantly more at the beginning of the observation period (24.64% in the morning and 12.06% in the afternoon). Between different observation periods, the frequency of lying, eating, drinking and sleeping behaviours were no different. Aggressive behaviour had a tendency to be related with the time of the day (0.23% in the morning and 0.51% in the afternoon). It was concluded that the cows learned to find and use the feeding and water quickly after the introduction to the new cubicle housing, but that lying places took approximately one month to be used by all cows.

Are cows motivated by being brushed?

D. Bizeray-Filoche[1], A. Prin[1], A. Feuvrier[1], M. Pierchon-Dujardin[2] and B.J. Lensink[3], [1]Institut Polytechnique LaSalle Beauvais, Equipe Comportement Animal et Systeme d Elevage, Rue Pierre Waguet BP 30313, 60026 Beauvais cedex, France, [2]De Laval France, BP55, 78340 Les Clayes sous Bois, France, [3]ISA, Equipe Comportement Animal et Systeme d Elevage, 48 bd Vauban, 59046 Lille cedex, France

This study was aimed at measuring cows' motivation for brushing after a deprivation period, by rewarding them or not after a milking session with access to a brush. Twenty nine cows were milked using an Automatic Milking System (AMS) placed in a Waiting Area (WA). After being milked, cows were allowed to go out to the WA through a selection gate, either to enter a Feeding Area (FA) or to enter a 10m² Brushing Area (BA), containing an automatic swinging rotating brush and a non-return gate towards FA. We tested whether being brushed or not influenced the time spent in WA and if a deprivation period had an effect on brushing behaviour. Cows were assigned randomly to 3 treatments (C: Control treatment, n=8; B: Brushing treatment, n=10; BD: Brushing Deprivation treatment, n=11) and studied for four weeks (w). In w0, all the cows were guided directly towards FA. In w1, B and BD cows were directed towards BA, and C cows towards FA. In w2, only B cows were directed towards BA. In w3, all the cows were directed towards BA, but the brush motor was turned off. Data collected on cow behaviour (time spent in WA and brushing behaviour) were analysed by SAS (GLM). B cows progressively increased time spent brushing (w1: 333±50s; w2: 431±67; w3: 534±80, $p<0.05$). In w1 and w2, no differences in behaviour were detected. In w3, C and BD cows spent less time in WA than B cows (C: 2702±374 s/milking/cow; BD: 2071±299; B: 3289±348; $p<0.05$). In w3, BD cows showed more brushing bouts/day than in w1 (w1: 1.6±0.1; w3: 1.9±0.1; $p<0.05$) and made contact more frequently with the brush than C and B cows (C and B: 1.5±0.1; $p<0.01$). This study showed that cows are motivated by being brushed, especially after a deprivation period.

An investigation of the fear responses of beef heifers

M. Mazurek[1,2], X. Boivin[3], B. Earley[1], M.A. Crowe[2] and M. McGee[1], [1]Teagasc, Grange Research Centre, Dunsany, Co. Meath, Ireland, [2]School of Agriculture, Food Science and Veterinary Medicine, UCD, Dublin, Ireland, [3]INRA, Clermont-Ferrand, Theix, URH-ACS, France

The objective was to investigate the fearfulness of heifers using four behavioural tests and to determine whether their fear responses were convergent, discriminant or stable. Sixty heifers (mean initial live weight 320 (s.d. 48.8) kg), comprising 42 purebred Simmental and 18 Simmental × Friesian-Holstein were used. Animal behaviour was investigated using 4 fear response tests: 1, flight test (day (d) 0 and 63); 2, docility test (d 2 and 65); 3, fear test (d 4 and 67); and 4, crush test (d 69). The flight test involved placing an animal singly in a 40m long outdoor holding area (i) recording the length of time (latency) for the single animal to locate the group of naïve animals at the end of the area and (ii) assessing the minimum distance that an animal permitted a human to approach. In the docility test, animals were placed singly in a 9 m^2 pen and after 30 s a person entered the pen. The time taken to move the animal into a 2.25 m^2 area (1.5 m square) located in the corner of the pen (30 s maximum) was recorded. The fear test (reactivity to humans) was performed using a $9 \text{ m} \times 4.5 \text{ m}$ pen with 1.5 m squares marked on the floor (18 squares in total). The crush test was performed during blood sampling and the number of head and feet movements, difficulty to enter the crush and difficulty to restrain the animal were recorded. Data that were not normally distributed were log transformed. Flight, docility and fear test data were subjected to multivariate principal component analysis (PCA) and crush test data to multiple correspondence analysis (MCA) using SPAD 6.5. No difference was found between the first and second flight test for latency time, however this test was negatively correlated with the approach distance (r=-0.35, P< 0.006). The minimum approach distance for the second flight test differed between genotypes, being shorter for the crossbreds (6.8 m, s.d. 3.9) than the purebreds (13.7 m, s.d. 7.8) (P=0.002). There was no effect (P>0.05) of genotype on any other recorded variables. The behavioural responses of the heifers using the flight, docility, fear and crush tests were condition specific and discriminant and no convergent data were found.

The effect of human approach direction and speed on bovine flight distance

M. Fukasawa[1], T. Kosako[1], D. Kohari[2], H. Tsukada[1] and K. Oikawa[2], [1]National Institute of Livestock and Grassland Science, 768, Senbonmatsu, Nasu-shiobara, Tochigi, 329-2793, Japan, [2]Ibaraki University, 4668-1, Ami, Ami, Inashiki, Ibaraki, 300-0331, Japan

Flight distance is often used as a reactivity measure within bovine behavioural studies. However, these studies may be non-comparable because of different directions and speeds of human approach producing different results. The aim of this study, therefore, was to determine the effect of approach direction and speed on bovine flight distance. Five Japanese Black breeding cows (44–45 mo; BW 506±18 kg) were used in the study and subjected to three approach directions at three speeds. The first approach direction was from the front (FRONT), the second from the right or left (SIDE), and the third from diagonally right or left from the rear (BACK). Speed of approach was 7km/hour, 4km/h and 1km/h and each level of approach direction and pace was allotted on each test day according to a 3×3 Latin-square design. Flight distance was measured five times according to each combination per cow where the order of measuring combination and measured cow were allocated at random in each day. One observer who was unfamiliar with test cows measured flight distance. The closest distance between an observer and a cow before it started to flee was defined as the flight distance. The random effect of animal and the fixed effects of the direction, pace and test day on flight distance were analysed. The flight distance in FRONT (1.92±0.16 m) was significantly shorter (p<0.05) than that in SIDE (2.45±0.16 m). The effect of the approach paces on the flight distance was also significant (p<0.001). The faster an observer approached to cow, the farther flight distance was. These results show that approach directions and paces to cattle markedly affect the flight distance. Therefore, the approach manner must be standardised for sharing or comparing flight distance measurement obtained by several researchers under various conditions.

Does overnight access to pasture reduce lameness in dairy cows?

N. Chapinal[1,2], A. Barrientos[1], C. Goldhawk[1], A.M. De Passille[2], M.A.G. Von Keyserlingk[1], D.W. Weary[1] and J. Rushen[2], [1]University of British Columbia, Animal Welfare Program, 2357 Main Mall, V6T 1Z4, Vancouver, Canada, [2]Agriculture and Agri-Food Canada, PO Box 1000, V0M 1A0, Agassiz, Canada

To reduce lameness in dairy cows, improved housing methods are required. Cows housed continuously on pasture show improved gait scores over those housed on concrete, but continuous access to pasture can compromise feed intake and milk production. We assessed the effect of housing cows on pasture overnight on hoof health and gait. Forty-seven Holstein cows were balanced for calving date, bodyweight, parity and initial gait score and semi-randomly assigned to either continuous housing in a free-stall barn or housing on pasture from 2000 to 0700 h and in a free-stall barn for the rest of the day. Treatments began 28 d before calving and ended 56 d after calving. Gait and hoof health were scored every 28 d. For 29 cows, dry matter intake (DMI) from TMR was monitored continuously. Analysis was done using mixed models. Gait score (1=not lame, 5=severely lame) did not differ between treatments (p=0.62) but primiparous cows had better scores than multiparous cows (1.99±0.11 vs. 2.47±0.06, p<0.001). Gait scores increased from 28 d before to 28 d after calving (increase=0.55±0.07, p<0.001), but no further changes occurred (p=0.37). Survival analysis showed that multiparous cows were more likely to develop a severe haemorrhage or ulcer (p=0.04), dermatitis (p=0.01) and severe heel horn erosion (p=0.08) than primiparous cows but there was no effect of treatment (p>0.10). There were no differences in DMI (control=15.18±0.88 kg/d, pasture=15.20±0.86 kg/d, p=0.98 over 56 d after calving) or milk production (control=32.06±1.19 kg/d, pasture=32.99±1.20 kg/d, p=0.56 over 21 d after calving). Parturition and parity are critical risk factors for lameness. Housing on pasture overnight is a practical way of increasing access to pasture without reducing feed intake or milk production but this exposure appears to provide no beneficial effects on lameness.

Estrus expression in spring calving dairy cows at pasture or in cubicle accommodation

J.F. Mee[1], G. Olmos[1], L. Boyle[1] and M. Palmer[2], [1]Moorepark Dairy Production Research Centre, Teagasc, Fermoy, Co. Cork, Ireland, [2]University of Edinburgh, Royal (Dick) School of Veterinary Studies, Summerhall, Edinburgh, EH9 1QH, United Kingdom

This study aimed to improve understanding of the reasons for lower fertility in cubicle-housed cows by comparing characteristics of standing estrus, as measured by radiotelemetry (HeatWatch) and visual observations, between Holstein-Friesian cows at pasture (PASTURE treatment) and those indoors in cubicle accommodation (HOUSED treatment). The study lasted 9 weeks from 26th March 2007 and used 46 spring-calving cows (mean calving date February 28th 2007) divided equally between the two treatments. Visual observations (20 minute sessions) of mounting and other sexual behaviours were carried out three times daily and cows were fitted with HeatWatch transponders. Milk progesterone (MP4) profiles were used to estimate dates of ovulation. The breeding season began on 23rd April. Data were analysed with SAS 9.1 software, using chi-square tests for categorical data and regression models for continuous data. Silent ovulation (ovulation indicated by MP4 profile but without any behavioural signs or HeatWatch mounts received) occurred in 8 HOUSED and 4 PASTURE cows ($p>0.05$). Sub-estrus events (ovulation indicated by MP4 profile but with only some behavioural signs and only one or two HeatWatch mounts received within 4 hours) occurred in more HOUSED (9) than PASTURE (3) cows (Chi-square=4.06, $p=0.044$). In contrast, standing estrus events (ovulation indicated by MP4 profile followed by three or more HeatWatch mounts received within 4 hours and/or the animal observed standing to be mounted) occurred in more PASTURE (21, 91%) than HOUSED (12, 52%) cows (Chi-square=8.68, $p=0.003$). When standing estrus occurred, there was no difference between treatments in duration of standing estrus (measured by HeatWatch) (PASTURE 5h 26min, HOUSED 4h 24min, $p>0.05$) or median number of mounts received during standing estrus (PASTURE 7.5, HOUSED 4.0, $p>0.05$). These results highlight biologically plausible reasons for lower fertility in cubicle-housed cows where estrus expression may be characterised as of low intensity, thus reducing the probability of conception.

Effect of group size on the consistency of movement-direction among cows in a grazing herd

S. Tada, M. Takahashi, K. Ueda, H. Nakatsuji and S. Kondo, Graduate School of Agriculture, Hokkaido University, Kita 9 Nishi 9 Kita-ku, 060-8589 Sapporo, Japan

For grazing herbivores foraging in a group, the consistency of movement-direction among herdmates would contribute to not only the group cohesion but also their foraging sites selection. This study investigated the effect of group size on the consistency of movement-direction among grazing cows by analysing their intra- and inter-patch movement behaviour. A total of 6 dry cows grazed on a pasture (mean sward height; 15.6cm) for 5hr/day were used. Experimental treatments were 3cows/1.6ha (GROUP-3) and 6cows/1.6ha (GROUP-6). During the first 1hr of grazing, positions of cows were recorded at 1sec interval by GPS, and their feeding station (FS) behaviour was observed. Patch was defined as the cluster of FSs. The threshold step number between FSs delimitating among patches was determined by the log-survivor analysis. For each intra- and inter-patch movement, the movement vectors of itself and the herdmates' center coordinate during the same time-period were generated. From these vectors, actual distance moved (ADM) and distance moved along consistent direction (DMCD: $ADM \times COS[\alpha]$, where α is the angular difference in movement-direction between the focal animal and herdmates) were calculated. The ratio of $\Sigma[DMCD]$ to $\Sigma[ADM]$ (DMCD-Ratio) was calculated. ANOVA (2 replications/treatment) with 3 individuals as block factor was performed. For intra-patch movement, ADM, DMCD and DMCD-Ratio were similar between GROUP-3 and GROUP-6 (13.7 vs. 13.2m, 8.1 vs. 9.1m and 56.4 vs. 69.5%, respectively). For inter-patch movement, both ADM and DMCD were shorter for GROUP-3 than GROUP-6 (5.9 vs. 13.6m and 2.0 vs. 10.9m, respectively) ($P<0.05$), and DMCD-Ratio was lower for GROUP-3 than GROUP-6 (26.5 vs. 79.8%) ($P<0.05$). The relatively large group size could enhance the consistency of inter-patch movement direction among grazing cows and this could be accompanied by an increase in inter-patch distance moved. Thus, group size would affect defoliation of pastures by changing not only grazing pressure but also grazing behaviour.

Effect of introduction of single and pairs of heifers into a herd of dairy cows on synchronicity at the feed rack

G. Neisen[1,2], B. Wechsler[1] and L. Gygax[1], [1]Federal Veterinary Office, Centre for Proper Housing of Ruminants and Pigs, ART, Tänikon, CH-8356 Ettenhausen, Switzerland, [2]University of Münster, Department of Behavioural Biology, Badestr. 13, D-48149 Münster, Germany

The study of social processes during the introduction of heifers into dairy cow herds helps to understand welfare problems caused by the confrontation with a novel social environment. We monitored the introduction of a single (singleton) and a pair of heifers (pair) on each of six dairy farms (herd size 22-44 animals). Using an automatic tracking system, we collected positions in space of each animal once per minute over six days each before and after introduction, 24 h/day. Here, we compare synchronicity (highest proportion of time that one animal spent along the feed rack with another) of 1) the singleton and the pair towards the cows, of 2) the cows towards singletons and pairs, within 3) the pair versus the pair with the cows. In addition, rates of agonistic behaviour towards the heifers were observed directly. Linear mixed effects models were applied. 1) Singletons reached higher synchronicity towards the cows of the herd than heifers of the pair (0.19 ± 0.2 vs. 0.16 ± 0.1, $F_{1;5}=7.33$, p=0.04). 2) Cows that had higher synchronicity with other cows before introduction reached higher synchronicity with the heifers. For a given synchronicity towards other cows before introduction, cows had higher values of synchronicity with the singletons than with pairs (Interaction: $F_{1;330}=6.14$, p=0.01). 3) The pair reached higher synchronicity with cows than among themselves (0.16 ± 0.01 vs. 0.11 ± 0.02, $F_{1;5}=23.14$, p=0.005). The singletons received more agonistic behaviour from the cows than the pair (8.4 ± 2.3 vs. 4.2 ± 0.6, $F_{1;11}=84.26$, p=0.033). Even though the pair had known each other before introduction, they reached higher synchronicity towards the cows than among themselves. However, singletons reached even higher synchronicity towards the cows. This high synchronicity most likely reflected proximity of animals exchanging agonistic interactions. Thus, it seems advisable to introduce pairs rather than single heifers.

Effects of age on heat pain threshold during castration of Holstein-Friesian calves

B. Earley[1], S.T.L. Ting[1,2], I. Veissier[3] and M.A. Crowe[2], [1]Teagasc, Grange, Beef Research Centre, Dunsany, Co. Meath, Ireland, [2]School of Agriculture, Food Science and Veterinary Medicine, UCD, Dublin, Ireland, [3]INRA, Clermont-Ferrand, URH-ACS, Theix, France

The objective was to investigate the effects of Burdizzo castration on the thermal nociception (stress-induced hypoalgesia) following Burdizzo castration in Holstein-Friesian bull calves of different ages. A CO_2 laser-based method (MPB Lamsor Inc., Dorval, QC, Canada) for measuring thermal nociception was used to detect the presence of castration stress-induced hypoalgesia (i.e. raised nociceptive threshold to pain) in calves. Sixty Holstein-Friesian bull calves were assigned to either burdizzo castration at one of the following age groups (n=10/group): 1.5, 2.5, 3.5, 4.5, and 5.5 months (mo) of age, or left intact at 5.5 mo of age. Blood samples were collected via jugular catheters on d 0 (day of treatment), on d 1 to 3, and by jugular venipuncture weekly from d 7 to 35 for plasma cortisol determinations. The integrated cortisol responses for the first 3 h after castration were threefold greater (P<0.001) in 5.5 mo-old castrates than controls, and the responses were reduced (P<0.03) by 46 and 35% in 1.5 and 4.5 mo-old castrates, but not (P>0.14) in 2.5 and 3.5 mo-old castrates. The laser beam was applied to the skin on the caudal aspect of the metatarsus –72 h before, and at 12, 24, and 48 h after castration. Thermal pain threshold was defined as the latency of the calves to react to the heat stimulus by performing hind-leg withdrawal behaviours. There was no treatment x time interaction on the thermal nociceptive threshold of the 5.5 mo-old calves. The thermal nociceptive threshold tended to be greater in the castrated than in intact animals (mean values for the main effect of treatment=10.0 and 7.7±0.85 s, respectively). Furthermore, threshold values increased from 12 to 48 h relative to baseline at –72 h (mean values for the main effect of time=9.1±0.80, 8.8±0.80, 10.0±1.01, and 7.4±0.18 s, respectively). There was an effect of treatment, time, and their interaction on the thermal nociceptive thresholds of the 1.5 to 5.5 mo-old castration groups. The thermal nociceptive thresholds of the 5.5 mo-old calves did not change (P>0.21) at any stage compared with calves between 2.5 to 4.5 mo-old. By contrast, the thermal nociceptive thresholds of the 1.5 mo-old calves were greater (P<0.0001) than older calves before castration at –72 h. It is concluded that castration stress and pain experienced by the 2.5 to 5.5 mo-old calves resulted in stress-induced hypoalgesia

Influence of artificial vs. mother-bonded rearing on sucking behaviour, health and weight in dairy calves

B.A. Roth[1], L. Gygax[2], K. Barth[3] and E. Hillmann[1], [1]ETH Zurich - Institute of Animal Sciences - Physiology and Behaviour, Universitaetstrasse 2, 8092 Zurich, Switzerland, [2]Federal Veterinary Office, Centre for Proper Housing of Ruminants and Pigs, ART Taenikon, 8356 Ettenhausen, Switzerland, [3]Institute of Organic Farming, Johann Heinrich von Thuenen-Institute, Research Institute for Rural Areas, Forestry and Fisheries, Trenthorst 32, 23847 Westerau, Germany

In artificial rearing, calves are often fed via an automatic milk feeder without the possibility to perform natural sucking behaviour. The majority of these calves show abnormal oral behaviours (e.g. mutual cross-sucking). Aim of the study was to investigate sucking behaviour, health state and weight gain of calves either reared artificially or with unrestricted and restricted contact to their mother, respectively. Two treatment groups of calves suckled by their mothers (A: twice daily for 15 minutes before milking, n=15; B: unrestricted contact, n=14) were compared to two control groups, both fed 8 L milk/day via an automatic milk feeder (A1: twice daily, n=14; B1 six times daily, n=14). Milk provision was stopped for all calves at 13 weeks of age. All calves were kept in the same barn and cows were milked twice daily. For each animal, frequency and duration of cross-sucking were recorded three times (four hours per day at two consecutive days, at the age of four, ten and fifteen weeks). Health state of each animal was assessed daily and veterinary treatments were counted. Animals were weighed weekly. For statistical analyses, linear mixed effects-models were used. Only one mother fed calf (A-calf) but 13 of 14 calves in both artificially reared groups performed cross-sucking (t55=-6.45, p<0.001). During the 15 minutes contact to their mothers, all A-calves were suckled by their dams. Health assessment revealed poorer health state of both mother fed groups (F1,51=4.19, p=0.046, mostly caused by diarrhoea), but counts of veterinary treatments did not differ. During milk feeding period, weight gain was better in mother fed calves (F1,54=60.24, p<0.001, 1.20 kg/d and 0.90 kg/d for mother fed and artificially reared calves, respectively). Calves that were allowed to perform natural sucking behaviour did not develop cross-sucking irrespective of whether they had restricted or unrestricted access to their mother.

Is the colostrum intake of newborn calves affected by time since birth and provision of heat?

E. Vasseur[1], J. Rushen[2], D. Pellerin[1] and A.M. De Passillé[2], [1]Laval University, Pavillon Paul Comtois, 4131, Québec (QC) G1K 7P4, Canada, [2]AAFC, 6947 #7 Highway PO Box 1000, Agassiz (BC) V0M 1A0, Canada

To ensure calf health, it is recommended in North America to feed 4L of colostrum before 6h after birth but dairy farmers often complain that calves have difficulties drinking this amount. A better understanding of factors affecting calves' motivation to ingest colostrum could help develop better management. Forty Holstein calves were tested to evaluate effects of time since birth (2 vs. 6h) and provision of a heat lamp on motivation to ingest colostrum in a 2 x 2 factorial model. Calves were randomly assigned to treatments balancing for birth weight and gender. Calves were provided colostrum (>68 g Ig/L) from a teat-bottle. 35% of calves drank 4L or more, 30% drank 3-4L and 35% drank 2-3L in a first meal. To ensure adequate intake of colostrum, we fed a second meal and tube fed colostrum at the end of the experiment. Weight was highly correlated with colostrum intake (r=0.56, p=0.01) and speed intake (r=0.42, p=0.02). There was a significant effect of gender on drinking duration (GLM p=0.01): females drank longer (average: 23±5.7 min) than males (average: 17±6.1 min). No significant effects (GLM p>0.1) of heat supply or time since birth were found on consumption. Provision of a heat lamp for 1h did not improve colostrum ingestion (p>0.1) but rectal temperature prior first meal was correlated with total intake (r=0.37, p=0.04). Time since birth did not affect colostrum intake. Although provision of the heat lamp did not increase colostrum intake, calves with a low rectal temperature ingested less colostrum, suggesting hypothermia as a cause of poor colostrum intake.

Optimal weaning strategies for high-milk-fed dairy calves

B. Sweeney[1], J. Rushen[2], D.M. Weary[3] and A.M. De Passille[2], [1]University of Edinburgh, Edinburgh, Scotland, United Kingdom, [2]Agriculture and Agri-Food Canada, PO 1000 6947 Highway 7, Agassiz V0M 1A0, Canada, [3]University of British Columbia, Main Mall, Vancouver, Canada

When calves are weaned off milk they lose weight and have poor grain intakes the week after. We compared gradual and abrupt weaning of calves fed 12L of milk/d by automated feeders. 40 calves were housed in groups of 4 and weaned at 41d either abruptly or with three gradual weaning strategies: over 4d, 10d or 22 d, with one calf within each group randomly allocated to each treatment, balancing for gender and birth weight. During the week when their milk allowance was first reduced, the 22d-gradually weaned calves had lower weight gains than the other calves (0.52 vs 0.97±0.12kg/d) (Mixed model: $P< 0.05$). During the week before weaning, calves on the 22d-gradual and 10d-gradual weaning made more visits to the milk feeder (23.5±2.1/d; 24.0±2.1/d) than the 4d-gradual and abruptly weaned calves (13.2±2.1/d; 9.5±2.1/d). All gradually weaned calves had lower weight gains (0.32±0.12kg/d; 0.14±0.12kg/d; -0.04±0.12kg/d vs 0.83±0.11 kg/d)($p<0.05$) and ate more grain prior to weaning than the abruptly weaned calves ($P<0.05$). During the week following weaning (d41 – 49), the 22d-gradual and 10d-gradual weaned calves made fewer visits to the milk feeder than the abruptly weaned calves (7.0±2.1/d; 11.1± 2.2/d vs 14.7± 2.1/d)($P<0.05$), and had better weight gains (0.50±0.12kg/d; 0.17±0.12kg/d) than the 4d- gradual and abruptly weaned calves (-0.15± 0.12; -0.26± 0.12kg/d)($P<0.05$). Frequent visits to the milk feeder can be interpreted as a sign of hunger and each weaning strategy resulted in calves being hungry at different periods. Gradual weaning improved grain intakes before weaning. However, gradual weaning that begins too early reduces milk intake but does not increase grain intake sufficiently to compensate. Abruptly weaned calves had the highest weight gains up to weaning but lost most weight after weaning. Overall, gradual weaning over 10d results in the least hunger and the best overall response to weaning.

The effect of floor type, social company or relocation on calves' sleep

L. Hänninen[1], A.M. De Passillé[2] and J. Rushen[2], [1]University of Helsinki, Research Centre for Animal Welfare, Fac. of Vet.Med., PO Box 57, 00014 University of Helsinki, Finland, [2]Agriculture and Agri-Food Canada, PO Box 1000, Route #7 Hyw, Agassiz, V0M 1A0, Canada

Sleep is essential for calf survival, but little is known of how production factors affect calves' sleep. To study how flooring, social company or relocation will affect calves´sleep, we kept 3-month-old Holstein bull calves (102±8kg) singly on bedded concrete floors (n=12, CONCRETE) or on rubber mats (n=11, MAT), or in pairs on concrete (n=12 pairs, PAIR). Calves were filmed for 24h (one calf per each PAIR), after which half were relocated to an identical, but unfamiliar room, and filmed for another 24h. We estimated daily duration, bout duration and frequency for sleep (S), rapid-eye-movement sleep (REMS) and non-REM sleep (NREMS) based on resting posture. The differences between treatments were compared with a repeated measures split-plot mixed model. CONCRETE-calves slept less than PAIR-calves but more than MAT-calves (3.4±0.5 h/d vs. 4.0±0.5 h/d p<0.06, or 2.9±0.5 h/d p=0.001, respectively). CONCRETE-calves had 55 ±11 S bouts, similar to PAIR calves´ (52 ±11) but more than MAT-calves´ bouts (41 ±11, p=0.03). CONCRETE-calves had 75±10 NREMS bouts, more than PAIR-calves´ (63±10) or MAT-calves´ bouts (62±10, p< 0.05). CONCRETE-calves´ NREMS bout length was 1.8±0.4 min, shorter than PAIR-calves´ (2.9±0.4 min, p=0.001) but similar to MAT-calves´ bouts. CONCRETE-calves had proportionally more REMS than PAIR-calves (p=0.04) but similar to MAT-calves (35.6±7.1% vs. 26.9±7.1% or 37.2±7.0%, respectively) Relocation increased calves NREMS (125.2±30.8 min vs. 155.1±30.6 min, p=0.02). Calves kept on concrete floor had more fragmented sleeping patterns, and longer sleep duration than calves on rubber mats. Social company decreased the proportion of REM sleep. Relocation to a new room decreased calves' NREM-sleep. Production environment and practices have an effect on calves´ sleep quality.

Heritability estimates of behavioural and physiological responses of Holstein Friesian heifer calves to a behavioural test

C.G. Van Reenen[1], J.T.N. Van Der Werf[1], B. Engel[1], J. Campion[1], C. Schrooten[2] and M.P.L. Calus[1], [1]Animal Sciences Group of Wageningen University and Research Centre (WUR), Edelhertweg 15, P.O. Box 65, 8200 AB, Lelystad, Netherlands, [2]Holland Genetics, P.O. Box 5073, 6802 EB, Arnhem, Netherlands

Marked consistency of individual differences over time has suggested that behavioural and physiological reponses of heifer calves to open field (OF) and novel object (NO) tests are mediated by stable underlying traits. We therefore examined a possible genetic basis of these responses. At the age of 9 months, Holstein Friesian heifer calves (n=311) with known pedigree (consisting of 2621 animals) were individually subjected to a combined test involving confinement for 3 min in a start box, and the ensuing exposure to an OF for 5 min, followed by the introduction of a NO that remained in place for 10 min. The following behavioural measures were obtained: the accumulated times spent in locomotion and the numbers of vocalisations during the OF and NO phases, and the latency to make contact and the accumulated time spent in contact with the NO. Heart rate (inter-beat intervals) was recorded continuously throughout the test (Polar heart rate monitor). The average heart rate (beats per min) and heart rate variability expressed as the root mean square of successive inter-beat interval differences (RMSSD) were calculated for the period of confinement in the startbox and for the OF and NO phases of the test. Heritabilities were estimated (ASREML) with a mixed model comprising random genetic animal effects, random contemporary group (batch) effects and fixed year-season effects. Initial heritability estimates were all greater than 0.2. The highest estimates (> 0.4) were observed for numbers of vocalisations, RMSSDs and the average heart rate in the start box. Although relatively high standard errors (between 0.1 and 0.2) warrant further research, these findings provide support for a genetic component in calf reactivity to acute stressors. The collection of data in additional calves, as well as the analysis of cortisol responses in conjunction with the other measures, are currently in progress.

Effect of lighting regime on the general activity of broiler chickens housed at two different stocking densities: preliminary results

A. Villagra[1], I. Olivas[1], E. Lazaro[2], M. Lainez[1] and A.G. Torres[2], [1]IVIA, Centro de Tecnologia Animal, Poligono La Esperanza 100, 12400 Segorbe, Castellon, Spain, [2]Universidad Politecnica de Valencia, Animal Science, Camino de Vera s/n, 46022 Valencia, Spain

One of the controlled parameters in broiler production is the light, both the photoperiod and the light intensity. Traditionally, photoperiods have been long (22-24 h light per day) but more recently, some findings about the convenience of reducing that photoperiod are being found. The aim of this work was to assess the possible positive effect of reducing the photoperiod on broiler chickens, housed at two different stocking densities. 448 one day old chicks were used in the experiment, and they were housed under two lighting regimes (continuous and decreasing from 23L:1D to 13L:11D) of an experimental shed. Sixteen pens were located in each room, 8 with 12 chicks/m^2 and 8 with 16 chicks/m^2 and their body weight and feed consumption were monitored weekly. Twelve of these pens were continuously recorded during five minutes each half an hour, for subsequent observation of the videos. Observations were carried out by counting the number of animals lying or standing in one photogram each half an hour, on alternate days from 10 until 44 days old. The proportion of animals performing each activity was analysed by logistic regression and density ($P<0.001$), age ($P<0.0001$), time of the day ($P<0.05$), lighting regime ($P<0.001$) and the interactions between these main factors were significant. In general, as stocking density increased, the proportion of birds lying was reduced (factor=0.0923), rounding 5%; the continuous lighting tended to increase the number of active broilers (factor=-0.4644) by 10%, as well as the age until 25 days old, when the proportion of active broilers decreased from more than 90% to 75%. Body weight and feed consumption were affected by the interaction of light and density ($P<0.0001$), although just 160 grams of difference could be reported. These results suggest that decreasing lighting programmes may improve general activity of broiler chickens, without jeopardising productive parameters.

Stocking density / group size effects on indicators of broiler chicken welfare

S.A.F. Buijs[1], F.A.M. Tuyttens[1] and L.J. Keeling[2], [1]Institute for Agricultural and Fisheries Research (ILVO), Animal sciences, Scheldeweg 68, 9000 Ghent, Belgium, [2]Swedish University of Agricultural Sciences, Department of animal environment and health, Box 7038, 750 07 Uppsala, Sweden

Stocking density effects on several indicators of broiler welfare were studied in groups of 8, 19, 29, 40, 45, 51, 61 and 72 animals, housed in 3.3 m^2 pens from 0 to 40 days of age (four replicates per group size). Behaviour was studied from week 2 to 6. Latency-to-lie (LTL, measuring leg strength) and tonic immobility (TI, measuring fearfulness) tests were performed in week 6. Data were analysed in a linear mixed model. Values presented are LSMEANS ± SE. Increasing group size (gs) led to decreased preening time (gs8: 26.5±3.8s, gs72: 12.3±3.8s, $F_{7,811}$=2.12, p=0.039), more sitting bouts (gs8: 1.9±0.1s, gs72: 2.5±0.1s, $F_{7,775}$=3.44, p=0.001) of shorter duration (gs8: 181±13s, gs72: 154±13s, $F_{7,798}$=2.38, p=0.021), shorter walking bouts (gs8: 3.2±0.3s, gs72: 2.0±0.3s, $F_{7,406}$=2.71, p=0.009) and more adjustments of the sitting posture (gs8: 2.2±0.2s, gs72: 4.4±0.3s, $F_{7,810}$=8.08, p=0.0001). LTL duration decreased with density (gs8: 326±37s, gs72: 174±37s, $F_{7,234}$=2.55, p=0.015) whereas TI duration tended to increase with density (gs8: 112±28s, gs72: 198±27s, $F_{7,236}$=1.86, p=0.078). However, deviations from these general patterns were common. For instance, TI duration was significantly longer (p=0.003) in gs72 (198±27s) than in gs51 (80±31s), but only tended to be longer (p=0.079) than in gs8 (112±28s). LTL duration decreased steadily from gs8 (326±37s) to gs45 (174±37s), but rose again towards gs61 (273±37s). Group size and density were confounded in this study, since we chose to keep total available area constant as this is what happens in commercial practice. The deviations from the general trends therefore may have been caused by increasing group size influencing the social structure for the lower densities, whilst in the higher densities physical space limitations became more important. While there is an overall trend for indicators to suggest poorer welfare at higher densities, even with this large number of group sizes no cut off point could be found.

RFID identification system to monitor individual outdoor use by laying hens

S.G. Gebhardt-Henrich[1], T. Buchwalder[1], E. Froehlich[1] and M. Gantner[2], [1]Federal Veterinary Office FVO, Centre for proper housing of poultry and rabbits, Burgerweg 22, CH-3052 Zollikofen, Switzerland, [2]Gantner Pigeon Systems GmbH, Montafonerstr. 8, A-6780 Schruns, Austria

Laying hens that are able to use an outdoor area are known to do this to varying degrees. One significant factor of the use of the outdoor run is flock size. In large flocks (about 10'000 hens) fewer than 5% of the animals are seen outside at the same time. So far it is unknown how long individual hens use the outdoor run. A system is needed which can monitor individual laying hens leaving or entering the house. Existing systems require substantial modifications of the pop-holes so that hens exit singly at a reduced speed. However, this might modify their behaviour of using the outdoor area. We employed the Gantner Pigeon System and monitored the individual movements of 34 laying hens with RFID tags (125 kHz) in a flock of 500. Eight 75 cm long antennae were placed in front and behind of the two pop-holes (width 70, 130 cm) at a distance of 16 cm. Up to 2463 changes between antennae per hen per day were registered. In 96.8% of the movements out and into the house the correct antennae were registered. 32 out of 34 tagged hens spent 3.6 hours (± 1.36) outside. Hens passed the narrow pop-hole significantly faster than the wider pop-hole (narrow: 0.76 s, wide: 1.14 s, Sign Rank Test: S=-141, P<0.004, N=31). The mode of the calculated speed of the tagged hens when passing the pop-holes was 1.5 m/s, but some hens reached more than 4.5 m/s. The results indicate that this RFID system could be used for the registration of fast moving animals like chickens without restricting their movements.

Behavioural reactions to open field and tonic immobility tests in growing broilers

L. Balazova and E. Baranyiova, University of Veterinary and Pharmaceutical Sciences, Veterinary Public Health and Toxicology, Palackeho 1/3, 612 42 Brno, Czech Republic

The objective was to explore the development and sex differences in habituation (by assesing decreasing excitability) of broiler chickens (n=24) by the open-field test (OF) and tonic immobility test (TI). Dynamics of behaviour changes within repetive OF allows to evaluate excitabilty of animals, while duration of imobilty in TI position indicates the intensity of fear response. Twelve males and 12 females were each exposed weekly to 10 min OF and TI (2 x 6 tests) from hatching until day 42 of life. The data were analysed with repeated measures ANOVA using Tukey post hoc test. The results showed that both sexes explored the new environment (OF) by locomotor activity, visual orientation and vocalisation. In males, frequency of all these behaviours significantly decreased (p<0.05) in 2nd OF trial and remained significantly attenuated till the end of the experiment (p<0.001, compared to 1st trial). In females, locomotor activity significantly declined between the 1st and 2nd OF (3.90±0.87 min vs. 0.24±0.14 min; p<0.001) and remained decreased until the last OF (p<0.001). Vocalisation frequency significantly declined in the 3rd OF trial (7.17±1.11 min vs. 1,00±0.39 min compared to1st trial; p<0.001) and remained attenuated also in the following OF trials (p<0.001, compared to 1st trial). Frequency of visual orientation was unchanged during the entire experiment in females. In all animals, the frequency of comfort behaviour progressively increased within the experiment (p<0.001). Time in TI increased significantly only in males (0.65±0.08 min vs. 3.66±0.66 min; 1st compared to last trial, p<0.01). These results show that during the early ontogeny males showed a lower level of excitability and their habituation was more obvious and progressive than in females. However, this was associated with a more pronounced fear response during later development.

Fear responses in different commercial mule ducks genotypes

I. Arnaud[1,2], E. Sauvage[2], E. Gardin[2], M.D. Bernadet[3] and D. Guémené[2], [1]SYSAAF, INRA Tours, 37380 Nouzilly, France, [2]INRA UR83, Unité de Recherches Avicoles, INRA Tours, 37380 Nouzilly, France, [3]INRA UE89, Unité Expérimentales des Palmipèdes à Foie Gras, INRA Artiguères, 40280 Benquet, France

Mule ducks, hybrids from the cross between a Muscovy drake and a Pekin female, are reported to express avoidance of people. In order to characterise effect of the genotype on fear responses, three parallel experiments have been conducted. Three pairs (one from the same company in each trial) of mule genotypes (60 ducks per genotype) with the same paternal or maternal origin, as well as their respective parental genotypes (30 ducks per genotype), were submitted to a set of different physiological and behavioural tests. Level of corticosterone in different conditions was measured. The tonic immobility duration was recorded. A corridor, in which other ducks or a man stand at one end, was also used to respectively assess the social motivation (Soc) or the response to man presence (RM). The duration of locomotion and the distance from stimulus were measured in these two conditions during the one minute of the test. Within each experiment, mule genotypes showed similar behavioural and physiological responses (p>0.05, for all traits). Thus, fear responses seem not to be influenced by genotype when mule ducks have a same paternal or maternal genetic origin. As regards to the comparison between mule and parental genotypes in the 6 crosses, adrenal response to acute stress and distance of avoidance of man were similar between mule ducks and their respective parental lines (p>0.05). However, mule genotypes expressed a higher duration of locomotion than parental genotypes during Soc tests (p<0.05 e.g. in one cross 44.8 for mule vs. 26.2 for Muscovy and 33.2 seconds for Pekin) and than paternal genotypes in RM test (p<0.05 e.g. in one cross 26.4 for mule vs. 11.8 seconds for Muscovy). Expression of panic behaviour in new environments in these hybrids, compared to the parental species, strongly suggest a higher emotivity background in mule ducks.

Modulation of D1 and D2 receptors affects food pecking by adult laying hens

R.T.S. Mc Gowan[1,2], J. Ware[1], K.R. Mc Cann[1], C.M. Ulibarri[1], S. Cloutier[1] and R.C. Newberry[1], [1]Center for the Study of Animal Well-being, Department of Animal Sciences and Department of VCAPP, Washington State University, Pullman, WA 99164-6351, USA, [2]Swedish University of Agricultural Sciences, Department of Animal Environment and Health, Section of Ethology and Welfare, Box 7038, 750 07 Uppsala, Sweden

We hypothesised that laying hens with elevated dopaminergic activity have an increased motivation to peck at chicken-flavoured food treats, a possible indicator of cannibalistic tendencies. To investigate the role of dopamine (DA) in the food seeking behaviour of laying hens, we manipulated DA activity in adult White Leghorns (n=60). We predicted that exogenous treatment with D2 agonists would promote pecking at chicken-flavoured treats whereas D2 antagonists would inhibit this behaviour relative to controls. We injected hens I.P with: D1AG-D1 agonist SKF82958, D1ANTAG-D1 antagonist SCH23390, D2AG-D2 agonist Quinpriole, D2ANTAG-D2 antagonist Raclopride or CONTROL-saline in two experiments (experiment 1 doses 0.1, 0.5, 0.5 and 0.1 mg/kg, respectively; experiment 2 dose 0.2 mg/kg for all; n=6 hens/treatment/experiment). Fifteen minutes after injection, individuals were given a choice between regular food pellets and a novel chicken-flavoured food treat for 5 minutes. Latency and number of pecks toward each option were measured. Drug treatment at dosages >0.1 mg/kg influenced overall food pecking frequency ($F_{4,25} \geq 7.38$, P<0.001), with D1ANTAG (76±17 pecks, 30±12 pecks) and D2AG (117±19 pecks, 15±9 pecks; for 0.2 and 0.5 mg/kg, respectively) birds pecking less than CONTROL birds (252±20 pecks; $t_{25} \geq 3.65$, P<0.01). Drug treatment at 0.5 mg/kg affected latency to peck at pellets ($F_{4,25}$=17.32, P<0.0001), with D1ANTAG (229±25 s) and D2AG (266±18 s) birds waiting longer to peck than CONTROL birds (49±15 s; $t_{25} \geq 4.93$, P<0.001). Whereas there was no effect on latency to peck the treat at the 0.1 or 0.5 mg/kg drug dosages ($F_{4,25}$=2.43, P>0.05), there was an effect at 0.2 mg/kg ($F_{4,25}$=7.99, P<0.001), with birds treated with D2ANTAG being quicker to peck the treat than birds on other treatments ($t_{25} \geq 3.74$, P<0.01). The results indicate that feeding was suppressed by the D1 antagonist and D2 agonist and suggest that, contrary to our prediction, cannibalistic tendencies may be heightened by activity of D2 antagonists.

The effects of quantity of reward on the choice behaviour of laying hens in a Y-maze preference test

S.M. Laine[1], G.M. Cronin[2], P.H. Hemsworth[1] and J.C. Petherick[3], [1]Animal Welfare Science Centre, Faculty of Land and Food Resources, University of Melbourne, Parkville, Vic, 3010, Australia, [2]Animal Welfare Science Centre, Department of Primary Industries (Vic.), 600 Sneydes Road, Werribee, Vic 3030, Australia, [3]Department of Primary Industries and Fisheries (Qld.), PO Box 6014, Rockhampton, Qld 4702, Australia

The quantity of reward in a preference test can be viewed as the time that the animal is allowed contact with its chosen resource. Effects of quantity of reward on choice behaviour of 15 laying hens (Hy-Line Brown laying strain) were examined. The birds were initially assessed in a runway test for their preference to be near either a familiar or unfamiliar hen. A familiarity score was given based on time spent near the familiar hen. Hens were then given a choice between dust (peat moss) or social contact (a familiar subordinate bird) in a Y-maze test for 8 trials, conducted on alternate days. Hens were randomly allocated to one of three quantities of dust reward (n=5): 'short' (2 minutes), 'intermediate' (20 minutes) and 'long' (45 minutes). If birds chose the social contact arm, they remained there for 5 minutes. When dust was chosen, the incidence of dustbathing was recorded. During testing, hens had social contact and dust in their home cage restricted. One-way analyses of variance, was used to examine treatment effects and familiarity score was used as a covariate. Treatment did not affect the choice made, however compared to other treatments, birds on the intermediate treatment tended to choose social contact more (proportion dust was chosen: 0.975, 0.825, 0.975 for short, intermediate and long respectively, P=0.07). There was a significant (P=0.014) negative relationship between familiarity score and the proportion of dust chosen trials in which birds dustbathed, indicating that a bird with a higher familiarity score was less likely to dustbathe when dust was chosen. Thus, preference for a familiar bird influenced how hens utilised the dust resource, which may potentially impact on choice behaviour. In conclusion, results indicate that quantity of reward may affect choice behaviour, however a more detailed investigation is warranted to validate these findings.

Statistical analysis of factors influencing the environmental choices of individual laying hens using T-maze tests

W.J. Browne, G. Caplen, J. Edgar and C.J. Nicol, University of Bristol, Clinical Veterinary Science, Langford House, Lower Langford, Bristol, BS40 5DU, United Kingdom

Preference tests have been influential in animal welfare assessment, although animals rarely make exclusive choices between environments. Understanding the factors that result in variation in choice is important for interpretation. We examined whether the responses of birds housed within different environments were significantly associated with their subsequent choices. We tested 60 laying hens (12 individual, 12 groups of 4) in three different environments (all 1.0 x 1.15m): wire floor (W), shavings floor (Sh) or shavings floor with additional peat, perch, and nest-box (PPN). The environments were experienced as three sets (W vs Sh; Sh vs PPN; W vs PPN) in a counterbalanced order. During housing periods (five weeks in first environment, five weeks in second environment, for each set) indicators of physical (e.g. plumage, body and foot condition), physiological (e.g serum cortiocsterone, H:L ratio and acute phase protein) and behavioural (e.g. tonic immobility, time budget, novel object) response were measured. After each set preferences between the two environments comprising that set were assessed with six T-maze choices per bird. Individual birds tended to make 'definite' choices at the end of each set but they frequently chose differently between sets with particular hens preferring different environments. Random effect logistic regression models explored the factors influencing choice while accounting for correlations due to the repeated choices of the birds and group housing. Relationships between the response measures were examined using a combination of cluster analysis and PCA. Factors associated with environment avoidance included higher fearfulness (tonic immobility OR 0.61 P=0.03), higher activity (frequency of movement between different pen regions OR 0.985 P<0.01) and higher head temperature (OR 0.85 P<0.01) in the environment subsequently avoided, compared with the same measures in the environment subsequently preferred. Experience of the environment immediately prior to test (OR 0.43 P<0.01) was also associated with subsequent avoidance. These models accounted for up to 80% of the variation in preference. By determining how measured responses influence animal choice, we provide a link between two very different methods of welfare assessment.

Is animal health status related to animal welfare: a case study with commercial broilers

L.R. Matthews[1], C.S. Bagshaw[2], A. Rogers[1], B. Jones[3], D. Marks[4] and A. Butterworth[5], [1]AgResearch, Animal Behaviour and Welfare, AgSystems, Private bag 3123, Hamilton, New Zealand, [2]Thinking Animals, PO Box 24089, Hamilton, New Zealand, [3]Inghams, Tower Rd, Matamata, New Zealand, [4]Livestock Solutions, PO Box 39715, Auckland 2145, New Zealand, [5]University of Bristol, Langford, Bristol, United Kingdom

Broilers in New Zealand (NZ) are raised using conditions and breeds typical of those in other countries. Comparatively, the NZ broiler industry has high-health (low-disease) status requiring low levels of vaccination. Lower levels of immune compromise imply a greater availability of nutrients for other functions. It was hypothesised that the high-health status would be reflected in better health/welfare as measured by incidence of mortalities, leg disorders and contact dermatitis. Mobility was recorded using a standard 6-point scale on 250-bird samples during visits to 36 farms over grow-out periods of 30-56 days, with maximum stocking densities between 33-39 kg/m2. Incidence and severity of foot-pad dermatitis and hock-burn were scored on samples of 100 chickens post-slaughter using standard 5-point and 4-point scales, respectively. Multiple regression models were used to determine the relationships between measures of welfare and management practices. Overall mortality was 3.8%; at the lower end of levels typically seen internationally (2.5-6.5%). The proportion of all birds culled for leg disorders was 0.3%. Importantly, the contribution of leg culls to total mortality was 8%, about half the levels reported internationally. Abnormal gait (score 3 or higher) was found in 23% of chickens; lower than that reported in another comparable international study (28%). The percentage with severe gait abnormalities (scores 4, 5) was 1.4%, which is 2 to 5 times lower than reported elsewhere. The proportion with moderate or severe foot pad dermatitis was 2.1%; at least 5 times lower than reported in other studies. Moderate and severe hock burn (scores 3, 4) were observed in 3.3% of birds, at the lower end of the international range (1-21%). There were no significant correlations between the welfare measures and stocking density. On most parameters, the welfare of NZ broilers was markedly better than that reported elsewhere; in support of the hypothesis.

Trade-offs between conspecific proximity or food against shock avoidance in fish

P.R. Laming, R.A. Dunlop and S. Millsopp, Queens University, Belfast, School of Biological Sciences, MBC, 97 Lisburn Road, Belfast, BT9 7BL, United Kingdom

Ethical considerations require an assessment of pain perception in farmed animals. To assess potential pain perception in fish, our laboratory has confirmed nociception in trout and goldfish. Subsequently, we have assessed escape responses to a potentially painful, discrete, 50 Hz, 3 or 30V electric shock delivered by a cable to the skin of the anterior flank when fish were in either of two stimulation zones of a four equal-zoned aquarium. MANOVA analysis showed that both goldfish (n =8) ($F_{2, 26}=36.1$; $p=0.0001$), and trout (n=8) ($F_{2, 20}=12.32$; $p<0.001$) avoided the stimulating zones and spend most of their time in the non-stimulating zones. In a second experiment, the high voltage zone was replaced by a zone containing a conspecific. Both goldfish ($F_{1, 13}=7.15$; $p=0.02$) and trout ($PF_{6, 5}=6.7$; $p<0.03$), spent more time in the adjacent low voltage shock zone than before. Recently, we have also performed 'trade off' experiments with goldfish trained to feed in one quarter of the aquarium where they were later to receive a shock. In the first experiment, shock intensity between 1V, 5V, 10V, 15V, 20V was the variable (n=5/group). In a second experiment, the variable was the number of days (0, 1, 2, 3) of food deprivation (n=5/group). Goldfish reduce the number of times they enter ($F_{4, 20}=12.4$, $p<0.001$) and the total time spent ($F_{4, 20}=6.7$; $p<0.001$) in the feeding/shock zone as shock intensity increases. Goldfish increase the number of times they enter ($F_{3, 16}=14.9$; $p<0.001$) and total time spent ($F_{3, 16}=16.4$; $p<0.001$) in the feeding/shock zone as food deprivation increases. They also increase the number of times they make tail-flip and escape responses. Thus, fish balance avoidance of a noxious stimulus against obtaining a desired resource (conspecific, food).

The effect of different closing rates of the pen side walls on pig behaviour

J.I. Song, H.C. Choi, B.S. Jeon and J.H. Jeon, National Institute of Animal Science, National Institute of Animal Science, R. D. A., Suwon, 441-350, Republic of Korea, 441-706, Korea, South

An experiment was conducted to examine environmental influences on the behavioural pattern of pigs. The resting areas of an enclosed growing-finishing pig house were checked in two seasonal ventilation systems, and the excretion habit of pigs influenced by the different closing rates (50, 75 and 100%) of side walls of pens was surveyed. 1. The excretion habit of pigs was not influenced by temperature, humidity and the flow speed of running air as they excreted in a fixed area of the side walls. However, the lighting effects on the excretion habit was observed because pigs excreted in the darkest area of the pig pen. 2. The accumulated height and width of faeces showed 10 and 30 cm; 5 and 25cm; and 3 and 20 cm for 50, 75 and 100% of closing rates of side walls, respectively. It indicates that pigs excrete all over the floor in the pen with 100% closed side walls. 3. Ammonia concentrations of the resting areas on the pen floor were determined to 4.2, 5.1 and 5.8 mg/l for 50, 75 and 100% of closing rates of side walls, respectively. It indicates that the ammonia concentration was highest in the pen with 100% closed side walls. Thus, the high ammonia concentration of the resting areas could be reduced by illuminating the darker areas with relation to the excretion habit. 4. The flow speed of running air was likely the biggest factor influencing the resting areas of pigs; pigs took a rest at the place of 0.03 m/s air flow speed point during midwinter, and at the place of 0.24 m/s air flow speed point during midsummer.

Behaviour and welfare of growing pigs in a "Bio-bed" group housing system

D. Kohari[1], Y. Matsumoto[1], K.I. Yayou[2] and S. Sato[3], [1]Ibaraki university, College of agriculture, Ami 4668-1 Inashiki-gun Ibaraki-ken, 3000331, Japan, [2]National Institute of Agrobiological Sciences, Kannondai 2-1-2 Tsukuba-shi Ibaraki-ken, 3058602, Japan, [3]Tohoku university, Graduate School of Agricultural Science, Amamiya1-1 Aoba-ku Sendai-shi Miyagi-ken, 9818555, Japan

The "Bio-bed" system, in which animals are reared on sawdust or rice hull floor with their fermented and decomposed faeces, is a popular system in Japan. We compared the behavioural and physiological characteristics, and responses to novel objects and humans between Bio-bed and conventional concrete slat housing system. Thirty of 400 pigs in each system were used for the study. Behavioural observations were made for 6 h from 9:00 to15:00 at 12 and 16 weeks of age. Maintenance behaviours (eating, resting, exploring, walking and others) were recorded every 40 sec using random sampling method. Blood samples for plasma cortisol measurement were collected from 10 of the 30 pigs after the behavioural observations. At 16 wk, a novel object (plastic cone) and humans (one woman and one man) were respectively presented to the 400 pigs in each system randomly five times for 30 sec after the behavioural observations. Latency to contact (LC), bite (LB), and the duration of biting (DB) were recorded. Maintenance behaviours in each system differed significantly in both periods (12 wk: χ^2_{4df}=40.9, P<0.0001; 16 wk: χ^2_{4df}=20.5, P<0.0005). In the Bio-bed system (24%), where pigs can perform rooting, exploratory behaviours were more frequent than in the conventional system (24 vs. 17%, re). Plasma cortisol levels were lower in the Bio-bed system than in the conventional system at 12 wk (t=1.65, P=0.058). In the behaviour for novel object, no difference was observed between the two systems. In the behaviour for humans, however, LC (woman: t=-3.48, P<0.01, man: t=-2.44, P<0.05) and LB (woman: t=-6.32, P<0.0005, man: t=-6.7, P<0.0001) were longer and DB was shorter (woman: t=7.47, P<0.0001, man: t=8.94, P<0.0001) in the Bio-bed system than in the conventional system. These results suggest that the Bio-bed system supports normal behaviour, moderates stress, and inhibit pigs' rough responses.

Methods of supplying chilled water for lactating sows during high ambient temperatures

J.H. Jeon, H.C. Choi, J.I. Song and B.S. Jeon, National Institute of Animal Science, National Institute of Animal Science, R. D. A., Suwon, 441-350, Republic of Korea, 441-706, Korea, South

The aim of this study was to determine an effective method of supplying chilled water (CW) to lactating sows during high ambient temperatures. One hundred twenty multiparous sows (Yorkshire × Landrace; parity range: 2 to 5) and their litters (Yorkshire × Landrace × Duroc) were tested under farm conditions with ambient temperatures above 25 °C. Four methods of supplying water were tested: control (free access to warm water at 22 °C), FACW (free access to chilled water at 15 °C), RACW (restricted access to chilled water at 15 °C), and RACW+SS (restricted access to chilled water at 15 °C with availability signaled using a sound cue). There were no differences among the treatment groups that received CW (P>0.05). Sows in all three CW groups ate and drank more than sows in the control group (P<0.01), and respiration rate and rectal temperature were lower in the CW sows (P<0.01). Litter size on 0 day postpartum and at weaning did not differ among treatments, but average weaning weight and average daily gain of piglets from the CW groups were higher than in the control group (P<0.01). These results indicate that access to CW benefits sows under high ambient temperatures, but that the three methods of supplying CW are similarly effective.

The relationship between the behaviour of sows at 6 months and their behaviour and performances at first farrowing

B.J. Lensink[1], H. Leruste[1], T. Leroux[2] and D. Bizeray-Filoche[3], [1]ISA Lille, CASE, 48 boulevard Vauban, 59046 Lille, France, [2]Pen AR Lan, route de Bovel, 35380 Maxent, France, [3]LaSalle Beauvais, CASE, 19, rue Pierre Waguet, 60026 Beauvais, France, Metropolitan

Piglet crushing remains a major problem in pig production. Reduced crushing might be obtained through genetic selection of behaviour. The aim of the study was to assess the relationship between behavioural responses at 6 months of age, around farrowing, and sows' reproductive performances including crushing levels. At 6 months, the behavioural responses of 75 nulliparous sows to humans were observed during a sow escape (of a human) test when in their home pen (scale from 1 – no fear to 5 – very fearful), and their responses to human presence in a weighing scale (scale from 1 – no agitated to 5 – very agitated). At first farrowing, the sows' nervousness when being in the farrowing crate one week before and the day of farrowing was observed (both on a scale from 1 – not nervous to 5 – very nervous), as well as the sows' responses when being approached at the front of the farrowing crate (withdrawal response from 0 to 10 sec.). The sows' escape response from the human at 6 months tended to be correlated with their response to presence in the weighing scale (r=0.21; P<0.10). The sows' escape response was also correlated with the withdrawal reaction when in the farrowing crate (r=0.38; P<0.01) and the sows' nervousness in the crate (r=0.24; P<0.05). The nervousness in the crate was significantly correlated with the nervousness around farrowing (r=0.34; P<0.01). The piglet crushing level was tended to be correlated with the escape behaviour of the sow at 6 months (r=0.21; P<0.10), and was correlated with the sows' nervousness at farrowing (r=0.28; P<0.05). These results suggest that behavioural responses to humans and management practices of gilts at 6 months of age are to some extend related with their behaviour around farrowing and crushing levels of piglets at first farrowing.

Aversion to pre-slaughter handling in five genetic lines of pig

A. Velarde, A. Dalmau, M. Font I Furnols, M. Gil and M. Gispert, IRTA, Animal Welfare, Finca Camps i Armet, Monells (Girona), 17121, Spain

In commercial abattoirs, pigs are exposed to different stressors before slaughter. The strength of the aversion and the animal reaction may be affected by genetic factors. The objective of the study was to compare the effect of genetics on behaviour during the pre-slaughter handling and the aversion to stunning with 90% CO_2. Three-hundred and fifty gilts, from five European breeding lines which were free of the RYR-1 gene, based on Pietrain, Large White (LW), and Landrace purebred lines, a Duroc synthetic line, and a Meishan (MS) × LW advanced intercross were tested individually, using aversion learning techniques and behavioural studies. The test facility consisted of a starting pen and a raceway that led to the crate of the dip-lift CO_2 stunning unit. The pig was allowed to cross the 412 cm length and 60 cm wide raceway for 60s. If after that time the pig was reluctant to move through the raceway, it was gently pushed into the crate. On closing the gate, the crate descended manually to the base of a 260 cm deep well that was prefilled with 90% CO_2. The time taken to enter the crate and the presence of aversive behaviours during the inhalation of the gas (retreat attempts, sniffing, and gasping and escape behaviour) were recorded and analysed by the Proc GLM and Proc GENMOD statements of SAS, respectively. MS x LW took more time to enter the crate compared with Landrace and Duroc (123.79±7.59 vs. 91.99±7.67 and 89.53±7.59 s, respectively; p<0.05). During CO_2 exposure, almost all pigs showed sniffing and gasping. Escape behaviours were performed by more Pietrains (97.18%) than MS x LW and LW (85.51 and 75.71% respectively; p<0.05). The results suggest that at the abattoirs, pigs need to be handled according to the genetic breed. During the CO_2 stunning, Pietrain appeared to react more adversely than other breeds.

Plasma cortisol and norepinephrine concentrations in pigs: automated sampling of freely moving pigs housed in the PigTurn® versus manually sampled and restrained pigs

D.L. Matthews[1], R. Poletto[2,3], J.N. Marchant-Forde[3], D. Mann[4], R. Guinn[4], S. Peters[4], J. Hampsch[4], Y. Zhu[4] and C. Kissinger[4], [1]Purdue University, Laboratory Animal Program, West Lafayette, IN, 47907, USA, [2]Purdue University, Department of Animal Sciences, West Lafayette, IN, 47907, USA, [3]USDA-ARS Livestock Behavior Research Unit, West Lafayette, IN, 47907, USA, [4]BioAnalytical Systems Inc., West Lafayette, IN, 47906, USA

Minimising the effects of restraint and human interaction on the endocrine physiology of animals is essential for collection of accurate physiological measurements. Our objective was to compare stress-induced cortisol (CORT) and norepinephrine (NE) responses in automated versus manual blood sampling in pigs. A total of 16 pigs (30kg) were assigned to either (i) automated blood sampling via an indwelling catheter using a novel penning system called PigTurn®, which detects the pig's rotational movement and responds by counter-rotating, allowing free movement while preventing catheter twisting; (ii) automated sampling while exposed to visual and auditory responses of manually sampled pigs; (iii) manual sampling by jugular venipuncture while pigs were restrained in dorsal recumbency. During sampling of (i), personnel were not permitted in the room; samplings of (ii) and (iii) were performed simultaneously in the same room. Blood samples were collected every 20 min for 120 min and measured for CORT (ng/mL) using mass spectrometry and NE (pg/mL) using HPLC. Effects of treatment and time were computed with mixed models adjusted by Tukey post-hoc test. CORT and NE concentrations were lowest in group (i) followed by group (ii), which were not different (CORT=24.8±5.9 and 32.1±7.2, NE=255.3±48.1 and 278.7±60.6; $p>0.1$). However, CORT and NE levels in manually sampled animals (iii) were highest (CORT=50.2±5.9, $F_{2,13}=4.74$ $p<0.05$; NE=747.0±59.0, $F_{2,11}=24.81$ $p<0.001$) compared to both automated methods (i and ii). Plasma concentrations across time were not different for CORT ($F_{6,71}=0.61$ $p>0.1$), but NE concentration at time 0 min (610.8±59.0) was higher than at 120 min (380.7±52.0, $F_{6,63}=3.54$ $p<0.01$). The presence of visual and auditory stimuli evoked by manual sampled animals did not affect non-handled pigs' responses. Restraint and manual sampling of pigs can be extremely stressful while, the automated blood sampling of freely moving pigs, housed in the PigTurn®, was significantly less stressful for the animals.

Sow responsiveness to human contacts and piglet vocalisation during 24 h after onset of parturition

H. Chaloupková[1], G. Illmann[1], L.J. Pedersen[2], J. Malmkvist[2] and M. Simeckova[3], [1]Research Institute of Animal Science, Ethology, Pratelstvi 815, 104 00 Prague, Czech Republic, [2]Danish Institute of Agricultural Sciences,Research Centre Foulum, Department of Animal Health, Postbox 50, 8830 Tjele, Denmark, [3]Research Institute of Animal Science, Biometric Unit, Pratelstvi 815, 104 00 Prague, Czech Republic

Sow responsiveness towards external disturbances and concurrent postural changes are proposed to be an important cause of early piglet crushing. The aim of the study was to assess whether Landrace–Yorkshire sows (n=17) change responsiveness over time within the 24 h after birth of first piglet (BFP) upon exposure to human contact and towards piglets' scream. The responsiveness was scored during: (i) blood sampling of the sow (BS), (ii) human handling of a piglet (HP), (iii) screaming when a piglet was trapped underneath the sow (SC) and (iv) exposure to playback of piglets' screams. A sow was scored as responsive if she changed her posture in response to the stimuli. The behavioural scores were analysed during: parturition (from BFP to birth of last piglet), phase 1 (from birth of last piglet to 12 h after BFP) and phase 2 (from 12 h after BFP to 24 h after BFP). During each phases a logistic regression model with mixed effects (proc GLIMMIX, SAS) was applied. The responsiveness towards BS was differed between the three time periods (L_2=9.7, p<0.01), sows were less responsive during phase 1 compared to parturition and phase 2. HP did not differ between periods (T_{16}=1.08, ns). We did not detect any changes in sow responsiveness towards SC (L_2=0.3, ns), which remained high (80%), whereas sows exposed to playback had a higher probability (T_{12}=0.20, p<0.05) to react at 12 h (50%) than at 24 h (25%). In conclusion, the responsiveness of sows toward direct human contact was lower during the first 12 h postpartum. However, sows were highly reactive towards the screaming of own trapped piglet during the whole 24 h postpartum. Decreasing responsiveness towards playback tests probably indicates a habituation of the sows.

Effects of infrared temperature on thermoregulatory behaviour in suckling piglets

G. Vasdal[1], E.F. Wheeler[2], A. Flø[3] and K.E. Boe[1], [1]Norwegian University of Life Sciences, Animal and Aquacultural Sciences, P.O Box 5003, 1432 Aas, Norway, [2]Pennsylvania State University, Agricultural and Biological Engineering, 228 Agricultural Engineering Bldg. University Park, PA 16802-1909, USA, [3]Norwegian University of Life Sciences, Mathematical Sciences and Technology, P.O. Box 5003, 1432 Aas, Norway

The objective of this study was to investigate the changes in thermoregulatory behaviour for suckling piglets at different infrared temperatures. Ten piglets for each of eight litters were at one week of age placed in an experimental creep area with either recommended infrared temperature (34 °C), 8 °C below (cool treatment) or 8 °C above (warm treatment). This procedure was repeated at 2 and 3 weeks of age week with recommended temperatures of 27 and 25 °C respectively. Digital photos were taken when all the 10 piglets had settled in the creep area, and the lying posture and huddling behaviour were analysed. A lying posture score and a huddling score was calculated by multiplying the number of piglets in each category with a given value for each category based on different lying postures and different degrees of huddling behaviour. The data were analysed using a mixed model analysis of variance with infrared temperature as a main effect and litter as a random effect. The piglets altered their lying posture significantly with different infrared temperatures in week 1 ($P<0.01$) and this effect was even more evident in week 2 ($P<0.001$) and week 3 ($P<0.001$). The proportion of piglets lying fully sternum decreased from 51% in the cool treatment to 25% in the warm treatment in week 1. Corresponding numbers for week 2 and 3 were 63% to 14% and 79% to 13%, respectively. The piglets preferred to huddle close together even at the higher temperatures in week 1, but in week 2 and 3 the huddling score decreased significantly with increasing temperature ($P<0.01$). This might suggest that while younger piglets tend to use huddling behaviour as a mean of thermoregulation, older piglets will rely more on changes in lying postures.

Influence of mixing on the initiation of piglet feeding behaviour and post weaning performance

F.H. Reynolds, J.M. Forbes and H.M. Miller, University of Leeds, Faculty of Biological Sciences, University Farms, LS2 9JT Leeds, United Kingdom

Weaning is a stressful period for the piglet, invariably typified by low voluntary feed intake and reduced performance. Moreover, if siblings are separated and mixed with other litters, stress increases and the learned stimulus of the group to feed is disturbed. This experiment investigated the difference between mixed litter origin or sibling groups on latency to initiate feeding and performance to day 6 post-weaning. Seventy-six piglets were weaned at 8.0kg ±0.28 (s.e.m.) and 28.1 ±0.39 days of age into 18 mixed sex groups with 4-5 pigs per pen. Piglets were allocated to either mixed (each individual from a different litter) or litter pens (siblings), balanced for weight and gender. Feeding behaviour was monitored using a multi-spaced FIRE system per pen providing ad-libitum access to feed (16.45 MJDE, 1.6g lysine/kg). Piglets were weighed at d0 and d6 post-weaning. Latency was defined as the time after entering the pen until a 0.5 minute feed. Data were analysed by pen using GLM procedures of Minitab 12.2; with replicate in the model and weaning weight as a covariate. There was no significant difference between the latency to first meal between mixed or litter treatments, with a mean latency of 379 min (±73.7). However, piglets in litter pens tended to have better performance than piglets in mixed pens ($P=0.068$) with average daily gains of 197 versus 77 g/d (±33.0) respectively. In conclusion, this study found no negative consequences of mixing unacquainted piglets on initiation of feeding post-weaning, refuting the hypothesis that separating litter-mates negatively affects feeding behaviour. Nonetheless, results suggest a post-weaning performance benefit when piglets remained with their littermates indicating that mixing piglets at weaning is disruptive. Further investigation into the contributory effect of group dynamics on piglet response to weaning is warranted to ensure optimal welfare and performance during this critical period.

Proximity seeking responses of piglets during separations and reunions

V. Colonnello[1,2], P. Iacobucci[1,2] and R.C. Newberry[1], [1]Washington State University, Center for the Study of Animal Well-being, P.O. Box 646351, Pullman, WA 99164-6351, USA, [2]University of Rome Sapienza, Faculty of Psychology, Rome, 00185, Italy

Our aim was to better understand social attachment in piglets and the impact of early weaning by investigating the seeking of proximity to familiar social stimuli in piglets exposed to either brief (Expt. 1) or long (Expt. 2) separations. In Expt. 1, we hypothesised that a brief reunion with mother and littermates would increase seeking of proximity to familiar social stimuli in re-isolated unweaned piglets. Between two 7 min-isolation sessions, 15-d-old piglets (n=16) were reunited with their mother and littermates in their home pen (Home group) or placed alone in an empty pen (Alone group) for 1 min. "Alone" piglets showed a greater reduction in calling frequency from the first (394±37) to the second (211±35) 7-min isolation than "Home" piglets (421±40 versus 363±33; repeated measures ANOVA: $F_{1,7}=19.49$, p=0.003; means comparison, p<0.01). During subsequent reunion, "Alone" piglets emitted more calls (13±2) than "Home" piglets (7±1; $F_{1,7}=6.89$, p=0.034), and spent more time near their mother's face (19±1 versus 11±1 s; $F_{1,7}=26.97$, p=0.001). In Expt. 2, we tested the hypothesis that seeking of proximity to the mother is preserved 48 h after weaning at 18 d. At 20 d, one weaned, and one unweaned, piglet from each of 8 litters was isolated for 7 min and then given a choice between the mother and an unfamiliar sow for 10 min. "Weaned" piglets emitted fewer calls (190±41) than "Unweaned" piglets (330±40) when isolated [$t_7=3.48$, p=0.01]. In the choice session, both "Weaned" and "Unweaned" piglets exhibited a comparable preference for their mother (p>0.05), spending more time close to her (326±38 s) than the stranger (161±30 s; $F_{1,7}=8.5$, p=0.02), and calling more frequently when close to her (464±70) than the stranger (239±57; $F_{1,7}=5.31$, p=0.05). These results suggest that the filial attachment bond is preserved for at least 48 h after weaning.

Effects of age at weaning on behaviour of piglets

N. Devillers and C. Farmer, Agriculture & Agri-Food Canada, Dairy and Swine R & D Centre, 2000 college St, PO Box 90, STN Lennoxville, Sherbrooke, QC, J1M 1Z3, Canada

Weaning is a disruptive process affecting social and feeding behaviour of piglets and age at weaning has been greatly reduced in recent years. The objective of this experiment was to study how weaning age affects piglet behaviour. Piglets were weaned and mixed at 21±1 (W21, n=81) or 43±1 (W43, n=70) days of age with 8 to 11 pigs/pen. Direct observations were performed 2 days before weaning (D-2), after weaning (D0) and 24 hours later (D1). Each piglet was observed by scan sampling every minute for one hour. Mean weaning weights were 6.3±0.3 kg and 15.3±0.5 kg for W21 and W43, respectively. The MIXED and the GLIMMIX procedures of SAS were used for repeated measures analyses with weaning age as the main factor. Time spent feeding on D0 was very low and increased on D1, being lower in W21 piglets (D0: 0.8 vs. 4.1% (P<0.001); D1: 9.5 vs. 12.9%, P<0.05; for W21 and W43 piglets, respectively). On D-2, W21 and W43 piglets rested for an equal amount of time (64.1 vs. 64.5%, P=0.93), whereas on D0, W21 piglets rested more than W43 piglets (83% vs. 61%, P<0.001), this effect being reversed on D1 (54 vs. 68%, P<0.001). Accordingly, W43 piglets increased time spent (P<0.01) in aggression, rooting and locomotion on D0 compared to D-2 and D1, while W21 piglets spent more time (P<0.01) performing these same behaviours on D1 than on D-2 or D0. In conclusion, weaning at 21 days induces a greater inhibition in feeding behaviour than weaning at 43 days. Timing of specific behaviours relative to weaning day differed between groups, W43 piglets displayed aggressive and exploratory behaviours on D0 with a reduction on D1, whereas W21 piglets were apathetic on D0 and displayed aggressive and exploratory behaviours on D1.

The presence of a buck reduces aggression in group-housed breeding rabbits, but may increase restlessness

S. Graf[1], L. Bigler[1], H. Würbel[2] and T. Buchwalder[1], [1]Centre for proper housing of poultry and rabbits (ZTHZ), Federal Veterinary Office FVO, Burgerweg 22, 3052 Zollikofen, Switzerland, [2]Division of Animal Welfare and Ethology, Department of Veterinary Medicine, Justus-Liebig-University of Giessen, Frankfurter Strasse 104, 35392 Giessen, Germany

Aggression in group-housed breeding rabbits (Oryctolagus cuniculi L.) may cause injuries and chronic stress, thereby compromising both animal welfare and production. However, group housing of breeding rabbits is recommended for better utilisation of space in larger enclosures and to enable the does to experience socio-positive interactions (e.g. allogrooming, resting in body contact). In practice, one buck is housed together with a group of females for about 10 days per reproduction period which is thought to have an appeasing effect on the does. The aim of the present study was to test this hypothesis by assessing the buck's influence on the does' activity and agonistic behaviour. Eight groups of five to seven 32 week old does (ZIKA-Hybrids) in their 4th reproduction phase were housed in littered pens (330cm x 175cm, 3 levels). General activity (feeding, locomotion, social interaction, grooming) and agonistic behaviour (aggression, fighting, fleeing) were recorded in each group from 24-hour videos one day before and one day after the buck was inserted into the group. The buck's effects on the does' general activity were assessed by scan-sampling at 4 minute intervals, while agonistic interactions between does were recorded by all-occurrence sampling. After the buck had been introduced, activity increased significantly from 40.67% per 24-hour (mean, SD=2.8) to 43.18% (SD=3.30; paired T-Test; N=8; T=-3.113; df=7; P=0.017), while agonistic interactions decreased from 7.75 (mean, SD=2.82) to 3.50 per 24-hour (SD=1.77; paired T-Test; N=8; T=3.440; df=7; P=0.011). These findings indicate that the presence of a buck in a group of breeding does can reduce potentially harmful agonistic interactions. However, further research is needed to assess whether the small but significant increase in general activity, which possibly reflects increased restlessness induced by the buck's mating behaviour, could have any adverse effects.

Effect of additional pen walls on resting behaviour in sheep

G.H.M. Jørgensen, I.L. Andersen and K.E. Bøe, Norwegian University of Life Sciences, Department of Animal and Aquacultural Sciences, P.O. box 5003, 1432 Ås, Norway

The aim of this experiment was to investigate if solid walls (partitions) in the resting area could decrease displacements from the resting area and possibly enable more sheep to rest simultaneously compared to a similar area without partitions. Six similar pens were constructed with concrete flooring in half the pen and a solid wooden resting area (1.5 x 2 m), giving a total area of 1.5 m^2 per animal and resting area of 0.75 m^2 per sheep (Council Regulation No. 1804/1999). In five of the pens additional pen walls were erected in the resting area in different arrangements, leaving one pen as a control. In front of the pen all sheep could eat hay ad libitum. A total of 24 pregnant, adult ewes of the Nor-X breed were randomly allotted into 6 different stable groups and individually marked. Each group (4 ewes) was introduced to one treatment pen for one week before 24 hour video recordings were made. All groups were observed in all treatment pens. We used instantaneous sampling every 10 minutes to score activity and resting behaviours while displacements were observed continuously over 6 hours during daytime (10 a.m. to 4 p.m.). General behaviours and displacements were analysed using GLM, while resting patterns were analysed using the Kruskal-Wallis test. There were no significant differences between treatments in time spent resting, feeding, standing/walking or resting in activity area (P>0.05). A tendency for more displacements being performed in some pens (e.g. cubicles: 5.4±0.9) compared to others (three walls: 36.2±0.6, P=0.062) was found, but most displacements were actually observed in the control pen (6.3±1.4). In conclusion, partitions dividing the original pen walls in the resting area into three tended to be beneficial in terms of reducing displacements from the resting area.

Effect of green odour on the feeding behaviour of sheep in a novel environment

S. Ito[1], M. Sutoh[2], E. Kasuya[3] and K. Yayou[3], [1]Tokai University, Department of Agriculture, Kawayo, Minamiaso, Aso, Kumamoto-ken, 869-1404, Japan, [2]National Institute of Livestock and Grassland Science, 2, Ikenodai, Tsukuba, Ibaraki, 305-8602, Japan, [3]National Institute of Agrobiological Science, 2, Ikenodai, Tsukuba, Ibaraki, 305-8602, Japan

The objective of this study was to evaluate effects of odour on sheep feeding behaviour in a novel situation. Sheep were fed in an open field test with or without green odour (trans-s-hexenal and cis-3-hexenol 1:1). Seventeen Corridale sheep were used. Animals were assigned to one of three odour treatments: (NO) Novel odour; the animals were exposed to green odour for the first time in the experimental period. (AO) Accustomed odour; the animals were exposed to green odour in their home pen at every feeding time (8:00-9:00 and 16:00-17:00) during a 3 week period, and (C) Control treatment; no odour was used. The experimental animal was introduced into an open-field with or without green odour for 30 min. The feed (300 g cutting of timothy grass) was presented 15 min after the onset of experiment. The animals were food deprived prior to the test. The duration (seconds) of specific behavioural categories (sniffing, standing, feeding and head on feed box without feeding) and the number of vocalisations were scored. There were no significant treatment differences in the behaviour during the first 15 min. After feed presentation, the animals in NO spent significantly less time feeding than in C (197.7 ±164.6 vs. 467.4 ±234.0 sec, $p=0.02$, ANOVA). The animals in NO spent more time sniffing than in C (216.3 ±194.4 vs. 19.0±30.9 sec, $p=0.026$) and in AO (47.4 ±81.5 sec, $p=0.052$). The number of vocalisations in NO (32.1 ±31.0) was significantly greater ($p=0.024$) than in C (0.6 ±0.9) and tended to be higher ($p=0.064$) in AO (7.0 ±12.3). These results suggest that providing a novel odour during feeding distracts from feeding behaviour in sheep in a the novel environment.

The use of infrared thermography as a non-invasive method to assess pain in sheep

S.M. Stubsjøen[1], A. Svarstad Flø[2], S. Larsen[1], P.S. Valle[1,3] and A.J. Zanella[1], [1]Norwegian School of Veterinary Science, Department of Production Animal Clinical Sciences, P.O. Box 8146 Dep., NO-0033 Oslo, Norway, [2]The Norwegian University of Life Sciences, Department of Mathematical Sciences and Technology, P.O. Box 5003, NO-1432 Ås, Norway, [3]Molde University College, P.O. Box 2110, NO-6402 Molde, Norway

The aim was to determine whether changes in eye temperature, measured using infrared thermography (IRT), can detect different levels of pain in sheep. Six ewes were included in a 3x3 Latin-square design study, and given the following treatments in a metal pen while remaining in visual contact with other ewes: 1) Noxious ischaemic stimulus by application of a forelimb tourniquet (S), 2) Noxious ischaemic stimulus and flunixin meglumine administered intravenously (S+F), 3) Flunixin meglumine (F). Maximum temperature of the curuncula lacrimalis was recorded for up to 60 minutes, including 15 minutes of baseline, 30 minutes during intervention and 15 minutes post-intervention. Behaviour was recorded during the same period. Mean time of tourniquet application was 13.5 min (SD=9-20). ANOVA with repeated measurement corrected for treatment sequences, and similar model for Latin-square design, was used for data analysis. A significant reduction in eye temperature, both from baseline to intervention and during the entire trial period, was detected for all treatments (p<0.01). The most marked change from baseline to intervention occurred in the S group (p<0.01). The least marked change occurred in the S+F group (p<0.01). No significant difference in eye temperature was found among the three treatment sequences during intervention (p=0.07). Interestingly, a significant reduction for both lip licking and vocalisation was observed between test periods 1 and 3 (p<0.05). Forward facing ears was the ear posture most frequently recorded over test periods and during treatments. A sympathetically-mediated reduction in blood flow and resulting decrease in heat loss from the orbital capillary beds in response to pain and stress may have caused the reduction in eye temperature. The limited number of animals observed and handling stress may explain the lack of significance between treatment sequences. However, the study indicates that IRT may be a useful, non-invasive method to measure pain in sheep.

Behavioural and physiological responses in sheep to 12, 30 and 48 h of road transport

D.M. Ferguson, D.D.O. Niemeyer, C. Lee, D.R. Paull, J.M. Lea, M.T. Reed and A.D. Fisher, CSIRO Livestock Industries, Locked Bag 1, Armidale NSW 2350, Australia

The aim was to determine the responses of healthy sheep to road transport under good conditions for 12, 30 or 48 h. Merino ewes (n=120; 46.9±0.39 kg) were allocated to road transport duration treatments with two replicates per treatment (n=20/replicate x transport duration treatment). Animal lying behaviour and vaginal temperature were measured during and after transport using data loggers. Blood and urine samples and liveweights were taken pre-transport, and 0, 24, 48 and 72 h after arrival. The data were analysed using analysis of variance and repeated measures analysis. There were no differences in lying times during the first 6 h after arrival. Sheep transported for 30 or 48 h lay down for longer than sheep transported for 12 h between 7 and 15 h after arrival, but this effect was reversed later during the 24 h after arrival. Increasing transport duration resulted in increased haemoconcentration, urine specific gravity and indicators of catabolism, but the levels did not exceed clinically normal ranges and the sheep generally recovered to pre-transport values within 72 h. Sheep transported for 30 and 48 h had lower liveweights on arrival (41.4 kg) than sheep transported for 12 h (44.4 kg) (P<0.001) but these differences were not evident 24 h post-transport. Temperature profiles indicated a rise during loading and the initial stages of transport, followed by a decline to more basal temperatures. Higher blood creatine kinase concentrations were evident on arrival after 48 h transport (193.3 U/L) compared to 12 h transport (155.2 U/L) (P<0.05). There were no effects of transport duration on plasma cortisol concentrations. These findings indicate that healthy adult sheep, transported under good conditions, can tolerate transport durations of up to 48 h, without undue compromise to their welfare.

Effect of prenatal malnutrition on the response to social isolation at six month of age in female goat kids

A. Fierros, R. Soto, M. Ramírez, A. Medrano and A. Terrazas, Universidad Nacional Autónoma de México, Secretaría de Posgrado, Departamento de Ciencias Pecuarias, Facultad de Estudios Superiores Cuauti., Km. 2.5 carretera Cuautitlán- Teoloyucan. San Sebastian Xhala, Cuautitlán Izcalli, 54714, Mexico

Goats are gregarious animals that respond strongly to social isolation. Few studies had been reviewing the factors during ontogenesis that affect the establishment of social bonding. Also we found that prenatal malnutrition impairs the mother-young bonding in this species. This study was conducted to assess whether prenatal malnutrition impairs the response to social isolation of female goat kids at six months of age. Two groups of female kids were used: 1) Control (C, n=10), from mothers fed with 100% of their nutritional requirements during pregnancy; 2) Underfed (UF, n=12), from goats fed only with the 70% of their requirements in protein and energy, from day 75 of pregnancy until birth. At six months of age, kids were tested regarding their response to social isolation. Each test lasted 10 min (5 min in presence of conspecific, 5 min in their absence). High-pitched bleats, locomotor activity, attempts and jumping out of the testing pen and eliminations were recorded. To compare between groups differences, the Mann Whitney U test was performed; to compare within groups, the Wilcoxon test was used. In both groups in all tests there was a significant increase of agitation following removal of the conspecifics (P<0.05). An agitation index was made in addition to other behaviours recorded in the presence of conspecifics and in their absence. Kids from UF group showed higher index of agitation in the presence of conspefics than C kids (3.48±0.51 vs. 2.46±0.4, P< 0.001). When conspecific were removed, C kids showed higher index of agitation than UF kids (4.5 ±1.6 vs. 3.2 ±1.2, P=0.04). Finally, the index of response to social isolation was higher in control than underfed kids (2.1 vs. 0.2, P=0.007). We conclude that prenatal malnutrition impairs the normal response to social isolation in female goat kids around puberty.

Effects of prenatal undernutrition in early gestation on the ewe-lamb bond and lamb behaviour in Suffolk and Scottish Blackface sheep

T.M. Coombs and C.M. Dwyer, SAC, Sustainable Livestock Systems, Kings Buildings, West Mains Road, EH9 3JG, Edinburgh, United Kingdom

In the UK seasonal differences in food availability may mean that extensively farmed sheep suffer from nutrient deprivation during early pregnancy. The aim of this experiment was to study the effects of early prenatal undernutrition on lamb behaviour and the ewe-lamb bond in two British breeds of sheep (Suffolk and Scottish Blackface). Control ewes (n=30) were fed 100% of requirements for maintenance and foetal growth throughout pregnancy (C) while restricted ewes (n=30) were fed 75% of requirements from days 1 to 90 of pregnancy and 100% thereafter (R). At 24 hours old, lambs were scored for reactivity during restraint and tested for their ability to recognise their own mother from another parturient ewe (alien) in a Y-maze. At 6-8 weeks old, data on lamb behaviour and ewe-lamb proximity in the field were collected by scan sampling. At 24 hours, Blackface R lambs struggled significantly more whilst restrained than Blackface C lambs (χ^2=6.892, p<0.05). During the maternal recognition test C lambs of both breeds were more likely than random to approach their own mother first (Blackface: χ^2=5.4; Suffolk: χ^2=8.33, p<0.05); the choices made by R lambs did not differ from chance (p>0.05). In addition, R lambs spent equal amounts of time with either their own mother or the alien ewe (p>0.05), suggesting that R lambs' ability to recognise their own mothers at 24 hours old was impaired. At 6-8 weeks old the ewe-lamb proximity in the field was significantly greater for Suffolk R lambs (n=17, median=47.5 m, w=179.5, p<0.05) than C lambs (n=15, median=6 m), and Blackface R lambs were seen lying more often and grazing less often (p<0.05) than Blackface C lambs. It is concluded that early prenatal undernutrition affects lamb behaviour in ways that may impact upon the ewe-lamb bond and lamb survival.

Development of behavioural scoring systems to assess birth difficulty and lamb vigour on farm

S.M. Matheson and C.M. Dwyer, Scottish Agricultural College, SAC, West Mains Road, Kings Buildings, Edinburgh, EH26 0PH, United Kingdom

Two welfare problems within the sheep production industry are the issues of low vigour lambs and high levels of birth intervention. Genetic solutions have the potential to help with these problems but need large amounts of data. Lambs that require assistance at birth are slower to perform neonatal behaviours, than unassisted lambs, and are less active for the first three days after birth. Lamb survival is dependent upon the expression of appropriate behaviours from both mother and offspring, and lamb vigour, i.e. the speed at which the lamb stands, finds the udder and sucks. However, there are many reasons why behavioural data are difficult to collect on farm, e.g. farmers are unskilled in behavioural data collection which can only occur at specific times of the year, at an already busy time; management systems are not amenable for data collection. Therefore the objective was to develop proxy methods (Scoring Systems) in order to assess the level of birth difficulties and lamb vigour on farm. Scoring data is relatively easier for the farmer to collect and has the potential to provide sufficient quantities of data. Data from over 1,000 lambs were assessed to create criteria for 4 Scores: Birth Assistance, ranging from unassisted short and long duration, minor and major assistance to veterinary assistance; Lamb Vigour (taken at 5 minutes of age), ranging from standing, attempting to stand, on chest to weak and lying flat; Sucking Assistance, which ranged from lamb sucking unaided within one or two hours to lamb requiring assistance for 1, 2 or 3 days post-birth; Lamb Mortality, ranging from healthy, live lamb to stillborn and born alive but dies later. These Scoring systems have been successfully trialled on farm with over 4,000 lamb records collected over the past 2 years and are being validated in two experimental flocks.

Behavioural effects of social mixing in feed-lot light lambs

G. Miranda-De La Lama[1], M. Villarroel[2] and G. María[1], [1]University of Zaragoza / Faculty of Veterinary Medicine, Department of Animal & Food Science, Miguel Servet 177, 50013 Zaragoza, Spain, [2]Politecnic University of Madrid, Department of Animal Science, Av. Complutense, s/n. Ciudad Universitaria, 28040 - Madrid, Spain

In Spain, lamb has traditionally been produced in extensive systems but this situation is changing toward more intensive production. The new scheme includes an intermediate step between the farm and the abattoir at cooperative classification centres (CC). Animals are loaded at the farm, unloaded at the CC, weighed and classified and distributed in feed lot pens until they reach the adequate slaughter weight. After that period they are reloaded to be transported to the abattoir. This process is a source of new stressors including social stress due to mixing. The aim of this experiment was to quantify the behavioural effects of social mixing at four different moments during the feeding process at the feed lot (days 1, 7, 14 and 28 at the CC). The experimental design included three replicates with 12 male Rasa Aragonesa lambs each (average weight at arrival 17 ± 1.5 kg and approximately 60 days old). Each group was observed using a continuous digital video camera system during 8 h per day (08:00–17:00 h) on each experimental day. The behaviours recorded as social (affiliation and aggressive) and oral stereotypes (biting, licking and chewing of through, water bucket and also bars of pen) were analysed using the minimum square techniques with SAS. Results indicate a significant effect ($p\leq0.001$) of social mixing in total aggression (38.25 ± 2.73; 28.14 ± 2.73; 22.29 ± 2.73 and 18.94 ± 2.73, on days 1, 7, 14 and 28, respectively). Aggressiveness was highest on day 1 and declined gradually. There was a significant difference ($p<0.001$) in affiliation behaviour, between day 1 (12.66 ± 1.07) and day 7 (7.27 ± 1.06), 14 (5.83 ± 1.06) or 28 (6.75 ± 1.06). On day 1, lambs presented significantly more ($p<0.001$) stereotypic behaviours (52.19 ± 3.22) than on day 7 (28 ± 3.22), 14 (25.61 ± 3.22) or 28 (21.29 ± 3.22). The results suggest that the first days in the CC are critical for animal welfare. It could be useful to provide environmental enrichment to minimise the effect of social mixing.

Influence of breed on behavioural and physiological responsiveness of lambs following gentling

M. Caroprese[1], A. Sevi[1], X. Boivin[2], G. Annicchiarico[3], T. Zezza[1], C. Tallet[4] and F. Napolitano[5], [1]Università di Foggia, Dipartimento PRIME, Via Napoli 25, 71100 Foggia, Italy, [2]URH-ACS, INRA, Clermont-Ferrand/Theix, 63122 Saint-Gènes-Champanelle, France, [3]CRA, Istituto Sperimentale per la Zootecnia, Via Napoli, 71020 Segezia (FG), Italy, [4] Institute of Animal Science, Ethology Group, Uhříněves, 10400 Prague, Czech Republic, [5]Università della Basilicata, Dipartimento di Scienze delle Produzioni Animali, Via Ateneo Lucano 10, 85100 Potenza, Italy

The interactions between genetics and human contact on later behaviour, stress physiology and immune responses was studied in eighteen Apulian (A) and eighteen Comisana (C) lambs allocated to two treatments: gentled (g) and not-gentled (ng). Treatment by trained stockmen consisted of gently stroking the lambs, 5 min each, three times a day for the first week of life and then twice a week for three weeks. In vivo cell mediated immune response to phytohaemagglutinin was evaluated at 7 and 28 d of age. At 30 and 31 d the lambs, in a randomised order, were subjected to two 5 min tests in a novel environment: a test of isolation and a test in the presence of the person who performed gentling. Blood samples were collected immediately before the tests, and 5 and 60 min after to evaluate cortisol concentrations. Variables were subjected to analysis of variance for repeated measures. When isolated, A lambs were more reactive in terms of number of bleats (57.38 ± 3.06 vs 41.76 ± 3.30, $F_{1,64}=12.02$, $P<0.001$) and climbing attempts than C animals (6.97 ± 1.46 vs 1.84 ± 1.57, $F_{1,64}=5.74$, $P<0.05$) without interaction effect. The number of contacts with humans were higher for Ag than Ang (13.78 ± 2.29 vs 4.00 ± 2.07, $F_{1,64}=3.70$, $P<0.05$), whereas no differences were observed between Cg and Cng. In the human test cortisol level of Ag was lower than Ang (53.88 ± 6.20 vs ± 5.61 ng/mL, $F=7.31$, $P<0.01$), whereas no effect of gentling was observed on C lambs. In the isolation test Ag showed a higher cortisol response than Ang (57.50 ± 6.20 vs 36.25 ± 5.61 ng/mL, $F=7.31$, $P<0.01$), whereas no effect of gentling was detected on Comisana lambs. No main or interactive effects of gentling or breed were observed on cellular immune response. Our results suggest that the effects of gentling on the responses of lambs to isolation or human presence depend on their breed.

Is the arena test a reliable method to assess sociability in lambs?

S. Ligout[1], J. Bouix[2], D. Foulquie[3] and A. Boissy[1], [1]INRA, UR1213 Herbivores - Adaptation et Comportements Sociaux, Saint-Genes-Champanelle, 63122, France, [2]INRA, UR631 - Amelioration Genetique des Animaux, Castanet-Tolosan, 31326, France, [3]INRA, UE321 La Fage, Roquefort-sur-soulzon, 12250, France

The present study aimed to validate an experimental approach to individually assess social reactivity among sheep. INRA401 male lambs (n=170) were reared together outdoors as part of a larger flock. Fifteen days after weaning the animals were individually exposed to an arena test carried out in a 7x2m enclosure surrounded by 2 m high solid walls. The test consisted of 3 phases: 1- social attraction toward 3 flock-mates placed behind a grid barrier at one end of the testing pen (30s), 2- social isolation (1min), 3- conflict between social attraction and human proximity (1min). Proximity toward conspecifics, vocal and locomotor reactivity were measured. The test period lasted 14 test days and 10 to 14 male lambs were tested each day (mean=12). At the end of each test day, tested animals were moved from the main flock to another pasture. The next day their inter-individual distances were measured when grazing over a 2-h period. This was done using scan sampling recording the identity of the nearest neighbour for each individual which led to the establishment of a sociability index. Overall, we found that high-pitched bleats recorded during the 3 phases of the arena test as well as the locomotor activity measured during the isolation phase were positively correlated with the sociability index (i.e. Pearson or Spearman correlations depending on the normality of the data). Correlation values with the sociability index ranged from 0.15 to 0.24 for high bleats measured during each phase (n=170, phases 1 and 2: $p<0.001$; phase 3: $p=0.05$) and 0.32 ($p<0.0001$) when extracting a principal component from those three measures. The correlation coefficient between the locomotor activity and the sociability index was 0.27 (n=163, $p<0.001$). The behavioural reactivity measured through the arena test thus reflects, at least to some extent, sheep sociability.

Do cats adapt to living indoors?

E.C. Jongman, Animal Welfare Science Centre, Department of Primary Industries Victoria, 600 Sneydes Road, Werribee 3030, Australia

Although most cat owners perceive that cats need to roam outdoors, cats generally appear to accept an indoor lifestyle, providing the indoor environment contains features that enable "natural" behaviour and redirect behaviour that could cause damage or could be classed as a nuisance to the cat owner. The aim of this study was to examine the relationship between undesirable behaviours and the level of confinement of domestic cats. A survey was conducted of 1600 cat owners recruited through eight participating councils within the Melbourne area, Australia, with a response rate of 38%. The survey included questions on demographics, general management, health, behaviour and behavioural problems. Cat owners were asked to rate the frequency of 29 'undesirable' behaviours of their cat on a scale from 1-5 (very often – never). Behaviours were classified into six behaviour groups (destructive, soiling, hunting, aggressive, nervous and fearful), and the owner's average score for all behaviours in a group was calculated. The relationship between behaviour group scores (appropriately transformed) and level of confinement was examined, using multiple regression analysis. Most of the behaviours on the survey were seen very infrequently, however the occurrence of these behaviours would usually indicate a problem for the cat or the owner. Cat owners either kept their cats 'confined at all times' (23%), allowed their cats outside for part of the time, varying from 'outside under supervision only' to 'free outdoor access' (69%) or kept their cats 'outside at all times' (8%). Owners reported that cats kept more indoors had more destructive (average score 4.4 (95% CI 4.3,4.6) vs 4.9 (4.8,5.0)), but less aggressive (4.8 (4.7, 4.9) vs 4.6 (4.4, 4.8)) and hunting behaviours (4.5 (4.3, 4.6) vs 4.0 (3.5, 4.3)), than cats kept more outdoors (P<0.001). However there were no relationships between confinement and soiling (4.8 (4.7, 4.9) vs 4.9 (4.8, 5.0)), nervous (4.5 (4.4, 4.7) vs 4.7 (4.6, 4.9)) and fearful behaviours (4.1 (3.8, 4.4) vs 4.2 (3.8, 4.5)) (P>0.05), suggesting that cats successfully adapt to their level of confinement.

A preliminary investigation into factors affecting aggression and urine marking in free-ranging dogs (*Canis familiaris*) in India

S.K. Pal, Katwa Bharati Bhaban, Life Science, Katwa Abasan, MIG (u)- 68, 713518, Burdwan, West Bengal, India

A preliminary investigation into factors affecting aggression and urine marking in free-ranging dogs (*Canis familiaris*) in India The purpose of this study was to investigate the influences of sex, season and places on the inter-group aggressive and urine marking behaviour of free-ranging dogs. Observation on the inter-group aggression and aggression related urine marking behaviour of 14 free-ranging dogs from two neighbouring groups (MIG-group: 3 males and 3 females, LIG-group: 5 males and 3 females) was recorded in this study. Animals were observed for 4 h /day; and each individual was observed for 102.86±7.17 (mean ± S.E.) h over 360 days. Behavioural data were collected using *ad libitum* and focal animal sampling. Mean (± S.E.) number of aggression for individual male was 249.38 (± 54.12) and that for individual female was 288.50 (± 66.17); and, the rate of aggression was a little higher among the females. Mean (± S.E.) number of urine marking for individual male was 120.88 (± 30.75) and that for individual female was 58.00 (± 10.05). So, the rate of marking was higher among the males than the females. In the case of male dogs, seasonal mean (± S.E.) number of aggression (112.38±26.95) as well as urine marking (59.00±15.66) was more intense during the late monsoon months, whereas, in the case of female dogs, seasonal mean (± S.E.) number of aggression (142.67±9.27) as well as urine marking (28.67±3.60) was maximum during the winter months. Frequency of aggression (χ^2=594.99, df=4, P<0.05) as well as urine marking (χ^2=478.92, df=4, P<0.05) varied with the places. Males were more aggressive and marked frequently in the mating places, whereas, females were more aggressive in the feeding places but marked frequently in the mating places. These results suggest that aggressive as well as urine marking behaviour in free-ranging dogs is influenced by sex, season and places.

Are problem behaviours in urban pet dogs related to inappropriate use of positive punishment?

C. Arhant[1], J. Troxler[1] and A. Mittmann[2], [1]Institute of Animal Husbandry and Animal Welfare, Department of Veterinary Public Health and Food Science, Veterinarplatz 1, 1210 Wien, Austria, [2]Institute of Animal Welfare, Ethology and Animal Hygiene, Faculty of Veterinary Medicine, Schwere-Reiter-Strasse 9, 80637 München, Germany

Effectiveness of positive punishment (inflict something unpleasant to reduce unwanted behaviour) is subject to conditions such as timing, schedule etc. Inappropriate use may impair dog welfare and is suspect of causing problem behaviour. Our aim was to quantify owners` use of positive punishment and to explore relationships of appropriateness of positive punishment with problem behaviour and training engagement. Therefore, we surveyed Viennese petdog-owners via questionnaire sent by post in January 2007. It contained questions regarding schedule, intensity and use of a warning signal in order to categorise owners according to the appropriateness of their use of positive punishment. Altogether 1345 questionnaires (return rate=28%) were analysed using Oneway-ANOVA, Cross-Tabulations and Linear Regression. 28.9% of the owners declared that they do not use positive punishment, 17.2% gave inconsistent answers and 43.6% constantly reported to use positive punishment. We concentrated on the latter in further analyses. Inappropriateness of positive punishment was linked with an increase in reported frequencies of aggression ($F(3, 506)=5.09$, $p=0.002$) and fear ($F(3, 509)=6.18$, $p=0.000$) in dogs. Moreover, decreased obedience ($F(3, 498)=36,80$, $p=0.000$) and a lower tolerance of close physical contact with the owner ($chi^2=22.91$, $p=0.000$) was found in inappropriately punished dogs. Frequency of dog training, a possible confounder, was related to appropriateness of punishment ($chi^2=30.17$, $p=0.003$) and dog behaviour: a higher frequency of dog training was associated with increased obedience ($F(4, 494)=11.87$, $p=0.000$) and decreased fear ($F(4, 505)=5.12$, $p=0.000$). Nevertheless, appropriateness of punishment showed to be the better predictor for obedience (beta=0.395, $t=9.78$, $p=0.001$) and fear (beta=-0.161, $t=-3.67$, $p=0.001$) in dogs. We conclude that inappropriate use of positive punishment has the ability to increase aggression and fear. It can lower the tolerance of close physical contact with the owner and decrease obedience. Therefore we see well-being of owners and public as well as dog welfare put at risk.

Effects of owner sex and interaction style on stress coping in human-dog dyads

I. Schöberl[1], B. Barbara[1], J. Dittami[1], E. Möstl[2], M. Wedl[1] and K. Kotrschal[1,3], [1]University of Vienna, Konrad-Lorenz-Research Station and Department of Behavioural Biology, Althanstrasse 14, 1090 Vienna, Austria, [2]University of Veterinary Medicine, Department of Biochemistry, Veterinärplatz 1, 1210 Vienna, Austria, [3]IEMT Austria, Margaretenstrasse 70, 1050 Vienna, Austria

If human-dog relationship is "truly" social, comparable with human-human dyads, there should be a distinct behavioural relationship between owners and their dogs and even between owner and dog stress coping. We predict that the interaction style of humans with their dogs as well as the sex of the owner will be contingent with dyadic stress coping. Our pilot study is based on 23 humans, 11 male and 12 female owners aged 23-68, with their medium to large sized intact male dogs from 1.5 to 6 years of age. During three sessions interactions of the owner-dog dyads were observed and video-taped in different test situations (e.g. mildly threatening of the dog by the experimenter moving toward the dog and gazing at it, owner present or absent). Saliva samples for the analysis of cortisol were taken simultaneously from owner and dog before, during and after the sessions and during two control days. Behaviour was coded from the videos with THE OBSERVER (Noldus). We found a positive correlation between the cortisol concentration of owners and their dogs after the threat situation (Pearson: n=20, r=0.495, p=0.027). Male dogs of female owners who showed less aggressive and less avoidance behaviour during the threat situation had higher cortisol concentrations thereafter than those reacting aggressively (Pearson: n=11, r=0.675, p=0.023). There was a positive correlation between the amount of interactions of female owners with their dogs during the threat situation and cortisol concentrations of their dogs afterwards (Pearson: n=12, r=0.601, p=0.039). Our results support the idea that stress coping of owners and dogs is related. Seemingly, owner interaction style and owner gender affect situational stress levels in the dog. In further steps we will also include owner attitude and dog personality.

Effects of an isolation period in an animal shelter on dog behaviour

L.M. Hemsworth[1] and J.L. Barnett[2], [1]Animal Welfare Science Centre, Monash University, P O Box 197, 3145, Caulfield East, Australia, [2]Animal Welfare Science Centre, The University of Melbourne, Royal Parade, 3010, Parkville, Australia

In animal shelters in Australia it is a legal requirement that all stray, abandoned or lost dogs undergo an initial 8-day isolation period, where individually housed dogs experience restricted human and canine contact, prior to consideration for re-housing. This study examined the changes in behavioural responses of dogs to humans and novel stimuli during the isolation period at a large Australian animal shelter. Forty-two stray or abandoned dogs varying in age, gender and breed were studied. Four 1-min behavioural tests were conducted on the dogs in their individual pens (approximately 1x 2.5 x 2.5 m, lxwxh) on day 1 (day 0=day of arrival) and day 7 of the 8 day isolation period: the four behavioural tests were response to a stationary human, response to an approaching human, response to a novel stimulus (coloured beach ball) and response to a startle stimulus (opening umbrella). The tests were conducted on individual dogs in the above order, with an interval of 2 min between each test. Data were analysed using Wilcoxon matched-pairs tests. There were no time effects on the behavioural response of dogs to the novel and starling stimuli. However, the dogs withdrew further in both human tests on day 7 than day 1 respectively (stationary human: 27.9 vs. 74.9 cm, P=0.006; approaching human: 42.7 vs. 101.8 cm, P=0.008). Furthermore, in the stationary test the dogs spent less time within 10 cm of the human on day 7 than day 1 (24.8 vs. 31.4 s, P=0.05). None of the behavioural variables were predictive of a dog's outcome in terms of euthanasia or re-housing. While these results are specific to the shelter studied, this study suggests a dog's fear response to humans increases during the 8-day isolation period. There is limited evidence in the literature that these behavioural changes may adversely affect dog welfare and clearly further research on this topic is warranted.

The calming effect of classical music on dogs: effects of music source and habituation

S.M. Rutter and L.H. Patrick, Harper Adams University College, Newport, Shropshire, TF10 8NB, United Kingdom

Previous studies have demonstrated that classical music has a calming effect on dog behaviour. Our study investigated two factors: 1) Whether the source of classical music (commercial classical radio station or a compilation of selected 'relaxing' classical music) influenced the calming effects 2) Whether dogs habituated to the calming effects of classical music. The study was carried out in a re-homing centre, with sixteen singly-housed dogs. Each dog received each of four treatments (in a Latin square design) for a period of six days: a) Commercial classical radio station (CR) b) Compilation of 'relaxing' classical music (CC) c) Commercial pop music radio station (PR) d) Control i.e. no music (NM) The dogs receiving each treatment could not hear the music from the other treatments. The dogs' vocalisation and activity behaviours were recorded using instantaneous sampling at 10 minute intervals for four hours on both the first and sixth (i.e. last) day of each treatment. Data were analysed using a Bonferroni test, with $p < 0.0083$ being considered statistically significant. The mean number of observations ($\pm$SEM) when the dogs were observed barking on day 1 and day 6 respectively for each treatment were: CR 4.1 ($\pm$0.98), 9.9 ($\pm$1.68); CC 3.8 ($\pm$1.31), 5.3 ($\pm$1.16); PR 11.9 ($\pm$1.82), 13.8 ($\pm$1.88); NM 12.3 ($\pm$1.73), 12.0($\pm$1.91). Dogs spent significantly less time barking during the two classical treatments compared with PR or NM on both day 1 and day 6 (the least statistically difference was between CR and NM on day 6; $t=-3.30$, $p=0.005$, with all other differences being $p<0.001$). There were no significant differences in barking ($t=-0.49$, $p=0.633$) between CR and CC on day 1, but there was significantly more barking ($t=4.07$, $p=0.001$) in CR than CC on day 6. Barking levels on day 6 compared with day 1 were significantly higher in CR ($t=-5.69$, $p<0.001$) but not significantly different in CC ($t=-2.52$, $p=0.023$). Similar results indicating that CC and CR were more calming than PR and NM were obtained for sitting and resting. The results support previous studies showing that classical music has a calming effect on dogs. However, this study shows that they habituate more quickly to a classical radio station than a 'relaxing' classical compilation.

Initial acceptance trial of flavours in a standard concentrate diet in horses at pasture

L. Williams[1], D. Goodwin[1] and P. Harris[2], [1]University of Southampton, Animal Behaviour Unit, University Road, Southampton, SO17 1BJ, United Kingdom, [2]Waltham Centre for Pet Nutrition, Equine Studies Group, Waltham on the Wolds, LE14 4RT, United Kingdom

Previous trials have focused on stabled horses in assessing flavour acceptance. In this trial we aimed to investigate flavour acceptance by pastured horses. The hypothesis was that horses would vary in acceptance of diets based on their flavour. Eight flavours used historically as flavourants were presented to eight horses maintained at pasture, at 1% in standard meals of 100 g cereal by-product. Diet acceptance, selection and consumption times were recorded. A Latin Square design was used to control for order effects. Before data collection, two meals of 100g of base diet (a standard cereal by-product) were offered, at the usual meal times, to approximate standard hunger levels. For the trial, horses were brought into individual familiar stables. Meals were offered in door mounted mangers, each used exclusively for that flavour to avoid residual effects. A minimum of one hour separated the meals which were terminated following complete consumption or no further interest for two minutes. Times for either; Complete consumption; Partial rejection (food partly consumed but then ignored for 2 minutes) or Total rejection were recorded. Any diet remaining was reweighed. Mean consumption times were compared using Kendall's coefficient of concordance. Variation in amount consumed was compared for all eight flavours using Kendall's coefficient of concordance. This showed a significant difference existed between amount consumed (W=0.314, n=8, p<0.01). Mean time to consumption or rejection was affected by individual variation and were non-significant for the group overall. The most accepted flavour by weight consumed was Agrimony (mean rank 2.81). The least preferred flavour was Elecampagne (mean rank 6.56). This trial showed flavour concentrations of 1% in a standard base diet significantly affected diet selection and acceptance, but not consumption times of pastured horses. These results may have practical implications in diet formulation for pastured horses.

Lying behaviour of horses kept in loose boxes and tie stalls: a preliminary study

M. Rundgren[1] and L. Kjellberg[2], [1]Swedish University of Agricultural Sciences, Dept. of Animal Nutrition and Management, P.O. Box 7024, SE-750 07, Sweden, [2]Ridskolan Strömsholm, Ridsportens Hus, SE-734 94 Strömsholm, Sweden

In Sweden most riding school horses are kept in tie stalls, which is debatable from an animal welfare point of view. The aim of this experiment was to compare the lying behaviour of horses kept in loose boxes with that of the same horses kept in tie stalls. This cross-over design was chosen as the number of horses was limited to 6 (3 geldings and 3 mares, age 5-15 years). Each horse was videotaped during 3 days in each of the two stable types, half of the horses started in a box and the other half in a tie stall. They were kept in the same system for at least four days before video-taping. All horses were used to both systems before the experiment started. The behaviours were manually registered from the tapes. Lying duration was registered continuously and measured in minutes and lying/getting up bouts was counted. The results were analysed with the statistical analysing system SAS, lying duration as a continuous variable in proc GLM with a model including system (box, tie stall), taping day nested within system (1, 2, and 3) and horse (1-6) as main factors. The number of lying and getting up bouts were analysed with chi^2 test. The total lying time and number of lying bouts per day differed between the individual horses ($p<0.001$), mean of horses: 74-215 min and 2.3-5.7 times, respectively. No significant differences were found between the two stable types, box vs. tie stall: 140 vs.129 min. and 3.9 vs. 4.0 bouts, respectively. However, in the box the horses performed a rolling movement immediately before getting up in 12 events out of 67. This behaviour was only observed once in the tie stalls (chi^2 $p<0.01$). The conclusion of the experiment is that tie stalls don't influence the lying duration of horses, but seems to suppress the rolling behaviour before getting up.

Does the daily level of voluntary activity influence the energy requirements of riding school horses?

H. Dekker[1], D. Marlin[1] and P. Harris[2], [1]Hartpury College, Hartpury House, Hartpury, Gloucester, Gl19 3BE, United Kingdom, [2]WALTHAM Centre of Pet Nutrition, Equine Studies Group, Freeby Lane, Waltham-on-the-Wolds, Leicestershire, LE14 4RT, United Kingdom

Critical to feeding for health and performance in horses is the adequate supply of energy. This study looked at the effect of voluntary activity (VA) on energy requirements in riding school horses. Twelve mature riding school horses of mixed breed were selected based on their bodyweight (BW) and estimated energy intake (EEI). EEI was based on the DE content of commercially available feeds and of hay. Horses were paired according to their BW, but within each pair one horse required a significantly ($P<0.01$) higher EEI (group H) than the other (group L) to maintain the same BW. VA was measured over 72hrs using two RT3 accelerometers positioned on an anti-cast roller and a head-collar. Objective visual observations were made using focal sampling. The yard staff scored the stable activity of the horses subjectively on an ordinal scale of 1 to 10. Differences in VA levels were determined using a Kruskal-Wallis test. Correlations between VA levels and horse characteristics were determined using Spearman's rho correlation coefficient (r_s). Roller (RA) and head-collar activity (HCA) were significantly ($P<0.001$) different between individuals. A strong correlation ($r_s=0.70$, $P<0.01$) was found between median RA and the perceived stable activity according to yard staff. Median RA and HCA were significantly ($P<0.05$) lower during the night (RA=3; HCA=36) than during the day (RA=18; HCA=335). Horses in group H and group L did not differ significantly ($P>0.05$) in age, BW, body condition, or workload. Median RA and HCA were significantly ($P<0.001$) higher in group H (RA=4; HCA=49) than in group L (RA=0; HCA=26) during the night period, but not during the day. Over the 72hr period, a trend ($P<0.10$) was found towards a higher VA level in group H (median RA=8; median HCA=191) than in group L (median RA=6; median HCA=149). Results indicate that daily VA levels may influence the energy requirements of horses with similar BW and management.

Differential reactivity to a novel stimulus in the equine oral stereotypy phenotype

A. Hemmings[1], J. Hayter[1], S. McBride[1] and C. Hale[2], [1]Royal Agricultural College, School of Agriculture, Cirencester, Gloucestershire, gl7 6js, United Kingdom, [2]Writtle College, Equine and Animal Sciences, Writtle, Essex, CM1 3RR, United Kingdom

Horses which perform oral stereotypy (crib-biting) exhibit alterations to nucleus accumbens (NA) physiology suggestive of sensitised dopamine transmission. In other species, enhanced dopamine transmission heightens the 'startle' response to novel stimuli; an effect that includes elevated heart rate (HR). In order to investigate similar phenomena in the horse we measured heart rate fluctuation in response to the presentation of a novel stimulus in crib-biting (n=7) versus control (n=7) populations. Within the confines of the animal's home stable a heart rate monitor was fitted for 30 minutes to yield mean basal heart rate (BHR). The novel stimulus (self inflating yellow rubber bag 40cm diameter) was then presented for 5 seconds. Heart rate was monitored for 30 minutes subsequent to stimulus presentation to measure stimulus heart rate (SHR). Mean SHR was then subtracted from BHR to obtain a figure equal to HR increase (HRI). In order to gauge the effect of stereotypy performance on HR, Average HR was recorded over a 30 minute period following feed induced stereotypy. The students t-test was applied to investigate differences in HR obtained from control and stereotypy populations whilst a paired t-test was performed to compare crib-biting versus basal HR values. All measurements are expressed as mean beats per minute (±SEM). Crib-biting and control animals displayed statistically similar BHR values (40±2.37 vs. 34.2±2.41, t=1.4 P>0.05). Moreover, there was no significant variation between basal and crib-biting HR (40±2.37 vs. 38±1.78 t=0.36 P>0.05). HRI as a result of stimulus presentation was significantly higher in crib-biting animals compared to controls (66±4.1 vs. 45±2.2 t=3.02 P<0.05). These results further characterise the behavioural effects of altered dopamine physiology as recorded in the equine oral stereotypy phenotype.

Behavioural and physiological assessment of stress level in hospitalised horses: correlation between parameters

M. Peeters[1], F. Péters[2], J. Sulon[3], C. Sandersen[2], P. Poncin[4], D. Serteyn[2] and M. Vandenheede[1], [1]Université de Liège - Faculté de Médecine Vétérinaire, Production Animale, Ethologie Vétérinaire et Bien-être des animaux - Boulevard de Colonster, 20, B43, 4000 Liège, Belgium, [2]Université de Liège - Faculté de Médecine Vétérinaire, Département clinique des animaux de compagnie et des équidés, Boulevard de Colonster, 20, B41, 4000 Liège, Belgium, [3]Université de Liège - Faculté de Médecine Vétérinaire, Département de sciences fonctionnelles, Physiologie de la reproduction - Boulevard de Colonster, 20, B41, 4000 Liège, Belgium, [4]Université de Liège - Faculté des Sciences, Biologie du comportement, Quai Van Beneden, 22, B I1, 4020, Liège, Belgium

Hospitalisation of highly reactive big animals, like horses, involves important stress reactions that complicate the medical care and disturb the healing process. The aims of this study were to assess stress levels in hospitalised horses, and to determine relationships between different physiological and behavioural stress measures, by using a Spearman correlation analysis. Thirteen horses requiring an elective surgery were housed in the Equine Clinic of Liège (Belgium) for 1 week. All measurements were collected during the 24 hours preceding surgery: plasmatic (PC) and salivary (SC) cortisol concentrations (sampling every two hours between 9:00 and 17:00); spontaneous behaviour (24h continuous video recording of the horses, alone in a box); behavioural reactions during a "novel object test (NOT)" and during a "handling test" (venous catheter placement); hearth rate (HR) before and during the handling test; horse's reactivity, assessed by the nursing staff (questionnaire, scales from 1 to 10). The mean plasmatic cortisol concentration was 50.3±10.3 ng/ml. The mean concentrations obtained for each horse, were significantly correlated with the: mean salivary cortisol concentrations (79.9±38.1 ng/100ml, r_s=0.88, $p<0.001$); standing duration in front of the box door (6.38±2.6 h/24h, r_s=0.72, $p<0.01$); moving duration during the venous catheter placement (5.42±7.03%, r_s=0.61, $p<0.05$); mean HR during resting time (36.8±5.7 beats/min, r_s=0.63, $p<0.05$); maximal HR during catheter placement (77.9±19.1 beats/min, r_s=0.76, $p<0.01$); touching frequency (2.8±2.0, r_s=-0.66, $p<0.05$) and duration (151±238 sec, r_s=-0.68, $p<0.05$) during NOT; excitability (4.9±2.0, r_s=0.65, $p<0.05$) and nervousness (5.0±1.4, r_s=0.65, $p<0.05$) evaluation (questionnaire). Interpretation of these variables in terms of stress will have to be confirmed by measurements in controlled stress conditions. This study puts the first bases leading to stress level appreciation of hospitalised horses.

The influence of coat colour and breeding index on fear reactions in Icelandic horses

E. Brunberg[1], S. Gille[1], G. Lindgren[2], S. Mikko[2] and L. Keeling[1], [1]Swedish University of Agricultural Sciences, Department of Animal Environment and Health, P.O Box 7038, SE-750 07 Uppsala, Sweden, [2]Swedish University of Agricultural Sciences, Department of Animal Breeding and Genetics, P.O Box 7023, SE-750 07 Uppsala, Sweden

The colour of a horse's coat has long been rumoured to reflect its temperament. But genes associated with colour often have additional biological functions and associations between coat colour and behaviour has earlier been reported in other species. Thus there may be some biological background to these claims. This study aimed to investigate if horses with mutations in PMEL17 (Silver dappled) or MC1R (Chestnuts) show stronger fear reactions than individuals without these mutations (Black or Brown horses). Twenty-seven Icelandic horses (Silver N=9; Chestnut N=9 and Black/Brown N=9 matched for sire) participated in the study. The horses were exposed to a fear stimulus (moving plastic bag) while eating from a bucket. The test was repeated five times and the reaction vigour was recorded on a six graded scale as well as reaction latency to return to the feed bucket after each trial. Handling difficulties to place the horse by the bucket before each trial were noted. The proportion of Silver horses that showed handling difficulties before each trial was higher (Wilcoxon; P=0.049) than in the other colours. No differences in reaction vigour or latency were found. However, offspring descending from two sires with a breeding index (BLUP) above 100 for a temperament trait 'Spirit' reacted stronger (Wilcoxon; P=0.01) and with a longer latency (Wilcoxon; P=0.004) than horses from the remaining two sires with an index below 100 for the same trait. It could not be shown that the mutations in PMEL17 or MC1R have an effect on fear reactions, but the differences in handling difficulties may indicate that the Silver horses are more suspicious in new situations. If the results that sires with high 'Spirit' BLUP index give more reactive offspring could be confirmed with a larger sample size, the selection on this trait should be re-evaluated.

Object permanence testing in the young horse

K. Uprichard, B. Hothersall and C.J. Nicol, University of Bristol, Animal Behaviour and Welfare, Langford House, Langford, BS40 5DU, United Kingdom

Object permanence is the notion that objects continue to exist even when out of sight. In this preliminary study we piloted three methods that could be used to examine the object permanence abilities of horses, using five foals. In Method 1 food was visibly displaced into one of two boxes to determine whether foals formed an association between the last sight of the food and its subsequent location after displacement. Object permanence would be demonstrated by initiation of search behaviour and selection of one box under conditions that controlled for olfactory, position and handler cues. On the first trial all foals selected the correct box, but the sample size was too small for statistical analysis. Thus, we analysed the first 3 (of 8) trials for each foal to minimise effects of learning based on prior success. Overall, a group score of 12/15 correct choices for the baited box was obtained (binomial test, P=0.017). In Method 2, a novel social stimulus (two goats) was occluded by a screen. Foals were expected to gaze at, or approach the pen containing the occluded goats under conditions that controlled for olfactory and auditory cues. The time spent gazing at the pen with goats (9.0%) did not differ significantly from the empty pen (13.8%), repeated measures ANOVA, and foals appeared fearful in this environment. In Method 3 foals observed goats walking towards and disappearing behind a long screen. Foals spent significantly longer looking at the area adjacent to the screen when goats were visible compared with control conditions (paired t-test=-6.698, p=0.001), but no extra time looking at this area after goats had disappeared (p>0.05).However, one foal convincingly tracked the goats' movements and continued to look at the point of disappearance for 20s. In conclusion, all foals initiated search behaviour in the food displacement task, with significant success in the initial trials. One of five foals tracked a disappearing social stimulus, suggesting that this method could be further refined to examine object memory in horses.

Policy Delphi methodology to evaluate perceptions of equine welfare in Ireland

J.A. Collins, A. Hanlon, S.J. More, P. Wall and V. Duggan, University College Dublin, School of Agriculture, Food Science & Veterinary Medicine, Belfield, Dublin 4, Ireland

Horses and the equine industry are important to both the Irish culture and economy. The majority of ethological studies on equine welfare have focussed on the horses' behavioural needs, not on human behaviour. Little is currently known of what industry perceives as acceptable, or what represents frequent practice regarding equine welfare. This study investigates these perceptions using a three round web-based Policy Delphi, an iterative qualitative method that seeks to generate the strongest possible opposing views on the potential resolutions of a major policy issue. Forty four individuals from key stakeholder groups (government, equestrian organisations, charitable bodies and industry sectors) were recruited. In Round One, their perception of potential welfare infringements was surveyed, using 29 vignettes. Response analysis identified 17 situations as having potential for welfare problems. Respondents were asked, in Round Two, to grade each situation on a nine point Likert scale (0=minimum importance; 8=maximum). Twelve mechanisms that could potentially be used to address equine welfare issues were also now presented for grading for likely efficacy, using the same scale. Median, 25th, and 75th percentile scores were determined. Round Two analysis showed that the greatest concern was expressed for welfare of horses at: unregulated sales/fairs (median score 7; 25th percentile 5, 75th percentile 7); unlicensed harness or ridden races (6; 4, 7); horse slaughter facilities (5; 3, 6); home – trading, dealers, sales preparation (4; 3, 6); during transport overseas (4; 2, 6), and that there was greatest confidence, as means to address equine welfare issues, in: social education programmes (5; 3, 7); equestrian organisations (5; 3, 7); specialist print media (5; 3, 6). These results will be used to rank problem situations and potential solutions for further study, in Round Three, of desirability, feasibility and methodology for redress. Policy Delphi has been a useful qualitative research technique to secure industry engagement, identify the main welfare issues and explore means of improving standards.

Recent advances in the application of mathematical methods to animal welfare

L. Asher[1], L.M. Collins[1], C.J. Nicol[2] and D.U. Pfeiffer[1], [1]Royal Veterinary College, Epidemiology, Northumberland Hall, Hawkshead lane, North Mymms, Hatfield, Hertfordshire, AL9 7TA, United Kingdom, [2]University of Bristol, School of Veterinary Science, Langford House, Langford, BS40 5DU, United Kingdom

Whilst the incorporation of advanced quantitative methods inspired by engineering and physical sciences has greatly advanced other areas of the life sciences, they have been under-utilised in the field of animal welfare. Exceptions are beginning to emerge and share a common motivation to quantify some 'hidden' aspect in the structure of the behaviour of an individual, or group of animals. The potential of pattern analyses, such as Markov chains, fractal analysis, and others, social network analysis and agent-based modelling for the analysis and interpretation of behavioural data in animal welfare science is discussed. Fractal analysis measures self-similar correlations within data, and has been successfully used as a marker of stress-related behaviour. Markov chains and time pattern approaches can be used to understand patterns in sequences of behaviour and may be particularly useful for objectively quantifying abnormal repetitive behaviour. Social network analysis is typically used for detailed study of the interactions and connections between individuals within a defined population, and could be used to understand how changes in housing affect agonistic behaviour. Agent-based modelling is a highly flexible approach, which can be used to first decompose a biological system into its constituent parts (for welfare science, this is likely to be a combination of individual behaviour and physiology, interactions between individuals and environmental factors), then run simulation models to make predictions about alternative scenarios. Recent models of group dynamics provide design principles for animal housing which could be implemented to improve welfare. We present real-life examples (primarily based on behavioural data from small groups of hens) to illustrate the function of these techniques, the types of data that they can be applied to, and their associations with other measures of animal welfare.

Nutritional up-regulation of brain serotonin induces abnormal self-injurious behaviour (Ulcerative Dermatitis) in C57BL/6J mice

B.D. Dufour[1], O. Adeola[1], H. Cheng[2], S.S. Donkin[1], E.A. Pajor[1] and J.P. Garner[1], [1]Purdue University, Animal Sciences, 125 S Russell ST, West Lafayette, IN 47907, USA, [2]USDA-ARS, Livestock Behavior Research Unit, 125 S Russell ST, West Lafayette, IN 47907, USA

Selective serotonergic reuptake inhibitors are the first line pharmacological treatment for compulsive behaviours in humans. Nutritional up-regulation of brain serotonin decreases compulsive behaviours in captive animals, such as feather pecking in hens and self-injurious behaviour in non-human primates. Increasing dietary tryptophan and carbohydrates each increases CNS serotonin. In this experiment we tested the hypothesis that nutritional up-regulation of serotonin would reduce compulsive hair-plucking behaviour in barbering mice. We used 16 cages of barbering C57BL/6J mice in a double-blind AB/BA crossover design (total duration of 24 weeks, 12 weeks on each arm). Mice were fed ad-lib, with one of two iso-energetic (3.9kcal/g) diets made from purified ingredients – a control diet equivalent to standard laboratory rodent diet (24% Protein, 10% Fat, 57.3% Carbohydrates, 0.24% Tryptophan), and a treatment diet with increased carbohydrates and tryptophan, and reduced protein (12% Protein, 10% Fat, 68% Carbohydrates, 1.02% Tryptophan). Treatment order was randomised and balanced with respect to sex and age. Barbering scores were taken every 4 weeks. Mice were sacrificed at 24 weeks, and brain catecholamine metabolism measured using HPLC. As predicted, the treatment diet increased brain serotonin metabolism ($F_{1,18}$=31.17, p=0.001). Contrary to predictions, the treatment diet elevated barbering scores in slight-to-moderately barbered mice by week 12, and had no effect on severely barbered mice ($F_{1,24}$=12.80, p=0.004). Unexpectedly, the treatment diet induced ulcerative dermatitis (UD) (Likelihood Ratio ChiSquare=10.58, p=0.0011) in the experimental mice. UD is characterised by open skin lesions and excessive scratching behaviour. 54% of the mice in the experiment developed UD; of these, 89% developed UD when on the treatment diet, and only 11% when on the control diet. Although the treatment diet here did not reduce barbering behaviour, it provides a tool to study the etiology and management of UD, both of which are currently unknown, to the detriment of C57BL/6J welfare.

Inhalant anaesthetics as an alternative to CO_2 euthanasia

I.J. Makowska[1] and D.M. Weary[2], [1]University of British Columbia, Animal Welfare Program, 2357 Main Mall, V6T 1Z4, Canada, [2]University of British Columbia, Animal Welfare Program, 2357 Main Mall, V6T 1Z4, Canada

Laboratory rodents are commonly euthanised using carbon dioxide (CO_2) gas, but recent studies have shown that rodents find this gas aversive. Anaesthetic gases are commonly used in surgeries to induce loss of consciousness and may be a suitable alternative to CO_2. The aim of our study was to use approach-avoidance testing to evaluate rat responses to isoflurane and halothane introduced at a flow rate of 4 L/min at concentrations that cause recumbency within 155, 135, 110 and 75 seconds. Rats (n=8) were trained to enter the bottom cage of a two-cage apparatus for a reward of 20 CheeriosTM. Isoflurane or halothane was turned on at the assigned concentration as soon as rats started eating, and rats were free to leave the test cage whenever they chose. The effect of treatment was analysed using a mixed model with rat specified as a random effect. On the first day of exposure to anaesthetics, rats remained in the test cage for a median of 56 seconds (range of 30 to 116 seconds) and 6 of the 8 rats became ataxic. On days 2-16, most rats left the test cage very quickly, remaining in the cage for a median of 7 seconds. Rats tended to stay longer with isoflurane ($F_{1,53}=4.04$, P <0.05) and with lower concentrations ($F_{1, 53}=4.15$, P=0.047). Ataxia was observed in only 19 out of 56 trials; 13 of these involved the one rat that consistently remained for an average (± SD) of 54±31 seconds regardless of treatment. We conclude that initial exposure to anaesthetics is only slightly aversive and that these agents could be a good alternative to CO_2 euthanasia. However, after one exposure most rats clearly avoided either gas, suggesting that repeated exposure to these anaesthetics is a welfare concern.

Behavioural and welfare impacts of reproductive management in a population of Eastern Grey Kangaroos (*Macropus giganteus*)

I.J. McDonald[1], A. Tribe[1], P. Murray[1], J. Hanger[2], B. Nottidge[2] and C.J.C. Phillips[3], [1]University of Queensland, School of Animal Studies, Gatton, Queensland, 4343, Australia, [2]Australian Wildlife Hospital, Beerwah, Queensland, 4519, Australia, [3]University of Queensland, Centre for Animal Welfare and Ethics, Gatton, Queensland, 4343, Australia

In rapidly developing urban districts of Australia, native wildlife can multiply beyond the capacity of the area. At the Pines Golf Course on the Gold Coast of Queensland, Australia, extensive housing and native habitat destruction has isolated a population of eastern grey kangaroos (*Macropus giganteus*). A population management plan (PMP) was devised for ecological sustainability and minimise risk of human-animal conflict. A combination of pharmacological and surgical methods was used to limit reproduction following chemical immobilisation. These included hormonal implantation (Suprelorin®) of 117 females (subadults and adults), surgical vasectomisation of 12 adult males (>50 kg) and 7 subadult males (5-50 kg), and surgical castration of 6 adult males (>50 kg) and 46 subadult (5-50 kg). Pouch young were left untreated. The combination was intended to give a high level of reproductive control beyond the lifetime of the Suprelorin® implants. Observations of untreated and treated kangaroos were made before (20 sessions), during (10 sessions) and after (20 sessions) the PMP at dawn and dusk for 1 hour. Agonistic (fighting, high walking, grunting and chest beating) and sexual (following/sniffing females and penis erection) behaviours were observed in male kangaroos. Individual behaviours such as grooming, scanning, locomotion, lying down and feeding were observed in all animals. Focal animal and scan sampling techniques were used. After implementation of the PMP a decrease ($P<0.05$) was seen in all sexual and agonistic behaviours of treated subadult males compared with untreated subadult males using a chi squared analysis. The following/sniffing behaviour was observed only once during 28 observations ($P<0.01$) of the castrated adult male kangaroos during focal animal sampling. Vasectomised adult males showed no significant changes in agonistic behaviours ($P>0.05$) compared with untreated adult males using a chi squared analysis. The following and sniffing behaviour was observed during 21 of the 29 observation sessions compared with 26 of the 28 observation sessions of untreated males. No stress related increases were seen in the individual behaviours of both male and females post treatment using a kruskal-wallis analysis. We conclude that the implementation of a PMP had limited impact on kangaroo's behaviour and there were beneficial reductions in sexual and agonistic behaviours, particularly in subadult males.

Behaviour and serum cortisol measurements in dolphins (*Tursiops truncatus*) kept in open and closed facilities

A. Luna[1], R. Valdez[2], M. Romano[2] and F. Galindo[1], [1]National University of Mexico, Faculty of Veterinary Medicine, Ciudad Universitaria, 04510, Mexico, [2]Centro de Investigaciones y Estudios Avanzados, Departamento de Fisiología, Biofísica y Neurociencias, Av. Instituto Politécnico Nacional 2508, 07000, México DF, Mexico

There is limited information on the behaviour of captive dolphins in open enclosures. The aim of this study was to compare individual time budgets, including swimming direction and bouts of locomotion, as well as serum cortisol levels of dolphins in open and closed facilities. Ten dolphins, six in open enclosures in the Yucatan peninsula and four in closed facilities in Mexico City, were observed during diurnal and nocturnal sessions. Focal sampling was used to record individual behaviours (i.e. locomotion, resting, events of spying, hitting, scratching and hops). In total, 36 and 48 hours were used respectively in open and closed pools. Overall, 24 blood samples were collected from the ten dolphins (15 and 9 samples respectively from open and closed facilities). Serum cortisol was determined using radioimmunoanalysis. Mann-Whitney and T-tests tests were used to compare behaviour and cortisol levels between enclosures. Principal Component Analysis (PCA) was used to summarise individual information and test possible relationships between behaviours and cortisol levels in the two types of enclosures. The frequency of locomotion bouts was higher in open enclosures (0.42/min vs. 0.26/min, U=2, n=10, p<0.05) however, the proportion of time swimming clockwise and resting was higher in closed pools (p<0.05). There were no differences in the proportion of time the dolphins swam counter-clockwise (p>0.05). The frequency of spying, hitting, scratching and hops were no different between enclosures (p>0.05). Cortisol ranges in open and closed facilities were respectively, 0.43 -1.05 µg/dL and 0.48 – 1.42 µg/dL, p>0.05). PCA revealed two factors; 'Intermittent swimming' and 'Continuous Swimming'. 'Intermittent swimming' was related to a significant negative load of serum cortisol levels (-0.76). When individual factor scores were compared between enclosures, a tendency for a higher load for 'intermittent swimming' was found in open enclosures (U=4, n=2, p<0.08). This study is useful to know more about the behaviour of captive dolphins in open enclosures and helps to put into context some of the behaviours shown in closed facilities.

Eliminative behaviour of high and low yield dairy cows at pasture

L.K. Whistance[1], L.A. Sinclair[2], D.R. Arney[3] and C.J.C. Phillips[4], [1]Faculty of Agricultural Sciences, University of Aarhus, Institute for Health, Welfare and Nutrition, Blichers Allé 20, P.O.Box 50, DK-8830 Tjele, Denmark, [2]Harper Adams University College, Department of Animal Production and Science, Edgmond, Newport, Shropshire TF10 8NB, United Kingdom, [3]University of Estonia, Estonian University of Life Sciences, Kreutzwaldi 46, Tartu, Estonia, [4]University of Queensland, School of Veterinary Sciences, Centre for Animal Welfare and Ethics, Gatton 4343, Queensland, Australia

It is not clearly understood whether cows help maintain good body hygiene by avoiding contamination with fresh faeces. The aim of this study was to increase understanding of eliminative behaviour of dairy cows at pasture. Twenty high- (GH) and twenty low yield (GL) Holstein-Friesian cows (T-test, P<0.001) were balanced for parity (T-test: P=0.9). Each cow in the two yield groups became the focal animal when about to defaecate. Data were gathered by manual observation over 24 hours between 10.00-17.00 and sequences of walking-(w), standing-(s) or lying-(l) were recorded before, during and after each event. Activity was also recorded: static=lying-(l), grazing-(g) and loafing, i.e., standing still-(lo): active=moving to different area of field-(mf), going to drink-(td) and catching up with herd-(cu). Uppercase letters denote walking or standing whilst defaecating: lowercase letters denote pre- and post-defaecation behaviour/activity. If the post-defaecation behaviour lasted for ≤10s, a second post-defaecation activity was recorded. Group results were analysed using the G-test. A total of 208 events were recorded for GH (mean 10.4, SD=2.2) and 230 for GL (mean 11.5, SD=2.0). Standing to defaecate then moving, involved 18/31 sequences and the most frequent were sSws/wSws (GH=0.26; GL=0.29) and lSw/lSws (GH=0.23; GL=0.20). The predominant activity sequence for both groups was gSg (GH=0.26, GL=0.26). Remaining lying to defaecate did not occur. Cows rose from lying to defaecate a total of 205 times of the 438 events recorded though low yield cows were more likely to express lSg (GH=0.13, GL=0.22; P=0.01), whereas high yield cows were more likely to express lSlo (GH=0.17, GL=0.08; P=0.03). In all recorded events, 383 stood and 54 walked whilst defaecating (P<0.001). Cows most often stood to defaecate when performing static activities before and after defaecating (P<0.001) but not when they were active before and after voiding (P=0.72). Standing to defaecate and moving forward was the predominant behaviour pattern. Walking whilst defaecating was most likely to occur when cows were simultaneously engaged in an 'active' activity.

Changes in feeding and standing behaviour identify cows at risk for calving difficulties

K. Proudfoot[1], J. Huzzey[2] and M. Von Keyserlingk[1], [1]University of British Columbia, Animal Welfare Program, 2357 Main Mall, Vancouver, B.C., V6T 1Z4, Canada, [2]Cornell University, 132 Morrison Hall, Ithaca, NY, 14853, USA

Dystocia can severely affect the health, reproduction, milk production and welfare of dairy cows. The aim of this study was to determine if pre-partum feeding and standing behaviours could identify cows at risk for dystocia. We followed the behaviour of 26 cows (parity=2.0, STD=1.2). Thirteen cows, with no other health problems, were classified as "dystocia" where at least two experienced farm workers were required to assist during the delivery. These cows were parity-matched with 13 "healthy" cows that required no assistance during parturition. Cows were video taped to determine the time of parturition and provided a new diet formulated for lactation immediately after calving. Cows were fed twice daily. Individual dry matter intake (DMI) and time spend feeding, standing and the number of standing bouts recorded 18hr before to 18hr after calving. A mixed model in SAS that included parity and dystocia as fixed effects was used to determine differences in feeding and standing behaviour between groups. Dystocia cows consumed 2.3 kg less and spent 33 min less time eating in the 18 hr before calving than healthy cows (4.4 ± 0.8 vs. 6.7 ± 0.8 kg/18hr, P=0.03; 55 ± 9.6 vs. 88 ± 9.8 min/18hr, P=0.02, respectively). Dystocia cows also had more standing bouts than healthy cows pre-calving (16 ± 1.5 vs. 12 ± 1.7, P=0.05). In the 18hr following calving, dystocia cows ate more than healthy cows (5.8 ± 1.2 vs. kg 2.6 ± 1.2 kg, P=0.04), and had a longer latency between their last meal before calving and parturition and were the most likely to consume a larger first meal ($R^2=0.38$). These results suggest that cows at risk for dystocia can be identified by changes in feeding and standing behaviour in the hours immediately before calving. Improved understanding of how dystocia affects behaviour will allow future research to identify management changes aimed at improving conditions for at-risk cows during the critical period around calving.

Influence of cubicle partitions on resting behaviour of dairy cows

B. Hörning[1], W. Linne[2] and M. Metzke[2], [1]University of Applied Sciences Eberswalde, Organic Animal Production, Friedrich-Ebert-Str. 28, 16225 Eberswalde, Germany, [2]University of Kassel, Nordbahnhofstr. 1 a, 37313 Witzenhausen, Germany

In modern dairy housing, cubicles with a metal framework are often used. These inelastic partitions could hinder the cows' movements while lying down, standing up or during lying. This study aimed to compare different cubicle types with side partitions which could move to the side. Two compartments each with 22 dairy cows (German Black Pied) were used. Four cubicle types were compared, a standard metal frame (type Dutch Comfort) and 3 types with flexible side partitions and flexible neck rails of 3 companies. Nine cubicles of each type were included in the study (4 - 5 per pen). Each compartment had 24 cubicles (2.5 x 1.2m, straw mattress). Cubicles located between 2 different types and last cubicles of a row were not included for the evaluation. Cows were classified into 3 size classes. 26 hours of direct observations were carried out by two observers simultaneously spread over 8 days. Standing up was observed 225 times and lying down 178 times. Postures while lying were recorded in 62 cases (30 min scans). Data were evaluated using Kruskal-Wallis- respective Chi-Square-Test. Cows in standard cubicles had significantly higher values for hitting cubicle partitions while lying down (38.0% of observations in this type), hitting neck rail while standing up (37.2%), repeated head lunging forward before standing up (15.8%), lying in contact with side partitions (45.4%), lying at the rear curb (28.5%). Smallest cows showed significantly less lying at the rear curb (0.1%) and lying in contact with side partitions (14.3%) as an average of all cubicle types. The aforementioned behavioural parameters indicate an impairment of the normal resting behaviour of cows in cubicles of Dutch Comfort type. Thus, cubicles with flexible partitions could be an alternative with regard to animal welfare.

Softer stall surface ameliorates the effects of reduced bedding on dairy cows' resting behaviour

M. Norring[1], E. Manninen[1], J. Rushen[2], A.M. De Passillé[2] and H. Saloniemi[1], [1]University of Helsinki, PL 57, 00014 University of Helsinki, Finland, [2]Pacific Agrifood Research Centre, Agriculture and Agri-Food Canada, PO 1000 6947 Highway 7, Agassiz, BC, V9M 1A0, Canada

Adequate resting is very important to dairy cows and lying time can be reduced if cows reject uncomfortable stall surface materials. Dairy farmers sometimes reduce the amount of bedding used and this can decrease the lying time if the cows reject the stalls possibly due to uncomfortable thermal properties or hardness of the stalls. We evaluated the effects of different stall surface materials and bedding amount on cows' activity and preferences in an unheated building. In experiment 1, 52 dairy cows were kept either on generously straw bedded concrete or unbedded sand stalls for al least 21 weeks. The daily lying time was longer for cows on straw bedded concrete than on sand (straw 749 ±16; sand 678 ±19 min/d; linear mixed model, P=0.001). In experiment 2, the lying time and preferences of 18 cows using 3 different stall surfaces (concrete, rubber mat, sand) were compared. All materials were lightly bedded with 2 kg of straw. In phase 1, the cows had access to stalls of one surface type and the lying time was observed over 6d. Lying times were longest on the mats (mat 796 ±13; concrete 742 ±13; sand 741 ±13 min/d; linear mixed model P<0.001). In phase 2, cows had access to two of the three types of stalls for 10d and their stall preference was measured. Cows preferred mat stalls to concrete stalls (77 ±5% vs. 23 ±6% of observations, ANOVA P<0.001) but showed no preferences for sand stalls versus either concrete or mat stalls. Abundant bedding in concrete stalls increased lying time compared to bare sand. However, when little bedding is available soft underlying materials are preferable. If bedding costs require limiting the amount of bedding used, rubber mats may help maintaining cow comfort.

Effect of social licking of different body regions on heart rate in dairy cows

B. Stockinger, S. Laister and C. Winckler, University of Natural Resources and Applied Life Sciences Vienna, Department of Sustainable Agricultural Systems, Gregor-Mendel-Strasse 33, 1180 Vienna, Austria

Social licking is a non-agonistic behaviour for which tension-reducing effects and reinforcement of social bonds have been described. Furthermore, behavioural reactions such as closing the eyes indicate a positive subjective state in the animals. The aim of this study was to investigate if social licking has a measurable influence on heart rate in both the actor and the receiver and if the intensity of this effect is dependent on the body region being licked. Licking bouts of up to 25 Simmental cows were recorded over a total of 18 days. Continuous observation was carried out for 6 to 7 hours per day from videos. Cows were kept in a sloped floor deep litter system with a permanently accessible outdoor loafing area. For all social licking interactions longer than 10sec (n_{actor}=249, $n_{receiver}$=280) the body region that was licked (head, ear, neck dorsal/lateral/ventral, withers and remaining body parts) was determined. Cardiac activity was recorded using beat-to-beat measurements. Data were analysed using linear mixed models. In actors, heart rate decreased during licking (90.4bpm) compared to the 5min pre-licking period (93.3bpm; $F_{2,700}$=6.58, p=0.002). This effect was more pronounced when licking the neck (89.4 vs. 93.8bpm; $F_{2,109}$=8.35, p=0.021) or the rest of the body (89.4 vs. 93.7bpm; $F_{2,189}$=10.12, p<0.001). There was no overall effect of social licking in the receiver. Heart rate only decreased after the rest of the body had been licked (87.0 vs. 89.1bpm; $F_{2,158}$=6.62, p=0.009). The present results indicate calming effects of social licking in the actor independently of the region licked. Although expected from behavioural signs, in the receiver such an effect was limited to parts of the body other than head, neck and withers. This might be due to the fact that categories of social licking (spontaneous, after solicitation, following agonistic interactions) have not been taken into account so far.

Effects of positive handling on behaviour and production of primiparous cows

D. Baroli, M. Minero, G. Vezzoli, D. Zucca and E. Canali, Sezione Zootecnica Veterinaria, Department of Animal Science, Celoria, 10, 20133 Milan, Italy

Dairy cows are in daily contact with humans, consequently positive interactions are a key management factor. The aim of this research was to assess the effect of gentle handling on human-animal relationships, behaviour at milking parlour and production of primiparous cows. Twenty Holstein Friesian pregnant heifers, of similar age and physiological state (two months before calving) were used: 10 heifers were positively handled for 3 minutes 3 days/week for 4 weeks. Ten control heifers did not receive the handling treatment and could not see or interact with the handled animals. All the subjects were tested by the forced approach (FAT) to a known and unknown person. The FAT tests were performed before gentle handling (T1) and 3 times after the end of the handling period (T2=one day after, T3=10 days after calving, T4=32 days after the end of the handling treatment). Avoidance distances to test person were recorded. The behaviour of all animals was observed twice after calving in the milking parlour. Milk production and the average milking speed were recorded for each subject. Before the gentle handling, the avoidance distances did not differ among the two groups of pregnant heifers and all the animals maintained a greater distance from the unknown than from the known person. Handled heifers showed lower avoidance distances both to known (mean 0.3 m vs 0.8 m) and unknown person (mean 0.8 m vs 1.3 m) than control heifers (Anova, F=4.01, P<0.01). Control subjects showed more kicks (mean 1.80 vs 1.08) (Mann-Whitney test P=0.10) and moved more during milking than handled animals (23% vs 4%) (Fisher test P<0.05). Milk production and average milking speed were not affected by treatment. Regular gentle handing could be useful in improving the ease with which dairy cows are milked and also seemed to improve the relationship between primiparous cows and humans.

Behaviour of dairy cows subjected to an aversive veterinary procedure

M.J. Hötzel, C.C.M. Gomes, L.C.P. Machado Filho, L.A. Teodoro and C. Gasperin, Laboratório de Etologia Aplicada, Dept. Zootecnia & Des. Rural, Rodovia Admar Gonzaga, 1346, Itacorubi; Florianópolis, Santa Catarina, 88.034-001, Brazil

In small dairy farms that lack appropriate handling facilities, cows might be restrained and subjected to veterinary inspection or treatment in their milking environment. The aim of this study was to assess the influence of repeated application of aversive unavoidable veterinary procedures in the milking environment on dairy cows' behaviour during milking and their reactivity to humans. A group of 17 Holstein cows in their second to fourth parity, kept on an intensive rotational pasture system and machine milked twice a day by two familiar handlers, were exposed to a thorough clinical examination for three consecutive days in the milking parlour immediately before milking. The procedure consisted of assessing rectal temperature; respiratory and cardiac frequency with a stethoscope; udder and lymph node palpation; thoracic percussion; detailed eye, mouth and tongue inspection; rectal palpation. Throughout the procedure, that lasted approximately 20 min, the cows were in pairs and one of the cows' handlers was present. Behavioural data before and after the procedure were analysed using paired t-test and repeated measures ANOVA. The behaviour of all the cows during the procedure indicated strong aversiveness. Treatment did not influence the flight distance (metres) kept from the veterinarian or from a person unknown by the cows, assessed on the pasture, before and after the procedure (veterinarian: before=1.2 ± 0.1; after 0.8 ± 0.2; unknown: 1.0 ± 0.2 after 1.2 ± 0.2; $p=0.3$), nor the number of agonistic interactions within the group observed before (7.1 ± 2) and after (11.5 ± 3) the procedure ($p=0.3$), or a score given to each animal to the ease and time to approach and enter the pen, avoidance behaviour to human approach and contact, frequency of tail and ear flicking, moving and kicking, urinating and defecating during milking, and the cows' response to a third person that approached and touched them during milking (before=0.3 ± 0.2; after 0.6 ± 0.1; $p=0.2$). These results do not support the conclusion that the repeated application of aversive unavoidable veterinary procedures in the milking environment will influence the behaviour of cows during milking or their reactivity to humans.

Influence of housing system on behavioural and immune responses of buffalo heifers

A. Bilancione[1], E. Sabia[2], G. De Rosa[1], F. Grasso[1], C. Pacelli[2] and F. Napolitano[2], [1] University of Naples, Dipartimento di Scienze del Suolo, della Pianta, dell'Ambiente e delle Produzioni animali, Via Università 133, 80055 Portici (NA), Italy, [2] University of Basilicata, Dipartimento di Scienze delle Produzioni animali, Via dell'Ateneo Lucano 10, 85100 Potenza, Italy

From November 2005 to October 2006 thirty-two buffalo heifers were used to evaluate the effect of rearing system on behavioural and immune responses. Sixteen heifers (group IR) were group-housed in indoor slatted floor pen with an outdoor paddock ($3.0+3.0$ m^2/animal). Sixteen others (group ER) grazed a fenced Mediterranean natural pasture of ~40ha. The animals at beginning of experiment were aged 8-9mo and weighed 190kg. Phytohaemagglutinin (PHA, 1 mg/animal; intradermal injection) was used to assess in vivo cellular immunity, whereas 5mg of keyhole limpet hemocyanin were injected subcutaneously to evaluate antibody response. Heifers at the end of the experiment were subjected to novel object test. Each animal was exposed to a novel environment (a 6x6-m paddock), where in the middle a traffic cone was present. Avoidance distance at manger was evaluated by an assessor walking (1 step/s) toward each animal until signs of withdrawal. The heifer was used as experimental unit. Antibody response data were analysed with ANOVA for repeated measures with housing system (non-repeated) and time and time x housing system (repeated) as factors. The other variables were subjected to ANOVA with one factor (housing system). Daily weight gain was higher in IR than ER animals (0.56 ± 0.02 vs 0.50 ± 0.02 kg/d; $P<0.01$), although puberty was reached at similar age (~670d). IR animals touched more (4.5 ± 0.9 vs 1.6 ± 0.3; $P<0.001$) and devoted more time to the traffic cone than ER animals (61.9 ± 14.5 vs 13.4 ± 4.7 s; $P<0.01$). Avoidance distance at manger was lower in ER animals than IR (0.54 ± 0.13 vs 1.36 ± 0.21 m; $P<0.001$). Skin thickness after PHA injection was higher in ER heifers than in IR (3.7 ± 0.3 vs 2.4 ± 0.3 mm; $P<0.001$), whereas the two groups showed a similar humoral immune response. In conclusion, ER heifers were less reactive to novel stimuli and developed a better relationship with humans and a higher immune responsiveness.

Are dominant cows the leaders of a beef cattle herd during grazing?

R. Šárová[1], M. Špinka[1], J.A. Panamá[1] and P. Šimeček[2], [1]Institute of Animal Science, Department of Ethology, Přátelství 815, 104 00, Czech Republic, [2]Institute of Animal Science, Biometric Unit, Přátelství 815, 104 00, Czech Republic

Ungulate herds may be led by specific animals but whether the leaders are dominant animals is not clear. We tested the hypothesis of dominant animals' leadership in a herd of 15 Gasconne cows (of known dominance status) with their calves on a pasture. The position of each adult cow was recorded with GPS in a one minute interval during 15 days in June 2005. To evaluate which animal was moving in the front of the herd we defined a position with respect to the herd's movement vector in the given time. This position was defined as normalised positions of perpendicular foots from animals to the line connecting herd centroids in the preceding and following minute. Because high correlation in time violates assumptions of the regression model we took just one randomly chosen observation from each hour (172 time points). These observations were averaged to give one "leader coefficient" for each cow. Based on speed of herd centroids, average speed of individuals, average distance between individuals, herd shape and proportion of individuals moving in the same direction as the herd, cluster analysis divided data points into two clusters: grazing (average speed of movement=7.4 m/min) and resting and slow movement (average speed of movement=1 m/min). Leader coefficients were correlated with dominance and we found that the dominant cows were leaders during grazing (were in the front of the herd) (Spearmann correlation, r_s=0.73, p=0.002) but their leading role was non-significant during slow movement and resting (Spearmann correlation, r_s=-0.25, p=0.36). However, even during grazing, the dominance explained only 4.7% of variability of leading at any specific moment (unlike the Spearman correlation where we compared the cows´ averages over the whole period). Thus, dominant cows have more influence on herd movement during grazing than subordinate ones, but quantitatively, this role is rather limited.

Use of GPS position and GPRS communication technologies to record spatial behaviour in extensive, aggresive Spanish cattle

M.E. Alonso[1], J.M. Lomillos[1], D. Bartolomé[2], J.J. García[2], M.A. Aparicio Tovar[3] and V.R. Gaudioso[1], [1]Veterinary Faculty. University of León, Dpto. Animal Production, Campus de Vegazana, s.n., 245071, Spain, [2]I.T.C.Y.L. Research and Thecnology Sub-Direction, Centro de Investigación del Toro de Lidia, Paseo de Canalejas, nº 77, 2º B, 37001, Spain, [3]Veterinary Faculty, University of Extremadura, Dpto. de Zootecnia, Avenida Universidad, sn, 10071, Spain

Spanish aggressive extensive cattle breeds are difficult to handle and typically respond to human proximity by stopping their behaviour and walking or running away from the observer. This makes it particularly difficult to perform behavioural studies on these cattle. The utilisation of GPS–GPRS technologies could allow us to study behaviour parameters without the presence of humans and in the long term improve animal welfare by developing handling practices more attuned to natural behaviour. The study examined the effect of relatedness (mother-daughter versus mother unrelated cow) on cattle spatial behaviour at pasture. Six GPS neck collars were fitted in 6 adult cows of 2 different cattle herds. In each herd, two collars were fitted to a mother and daughter and another collar was fitted to a unrelated cow to compare the distances between related and unrelated animals. The GPS collars sent position information by GPRS to a web farm management service that made data from each animal instantaneously accessible. Location coordinates of the study animals over 1 month were transformed into distances between animals using Arc Map 9.3 program and Excel 2007. Data were analysed by one-way ANOVA to assess the significance of differences between pairs of related and unrelated cows. The mean distances between mother and daughter (1) were 283 (±139 SD) m and 182 m (±177SD) in each herd, and the distances mother-unrelated cow (2) and daughter-unrelated cow (3) were 164 (±126SD) and 255 (±180SD) m and 338 (±172SD) and 135 (±110SD) m, respectively. There were significant differences ($F_{(2, 1167)}$=143,8 p<0.01 and $F_{(2, 1397)}$=66,07 p<0.001) between the mean distances of the three pairs considered in each herd, but these were not consistent in direction. The high distances observed between the No. 1 pair of both herds, being not significantly different from the other two possibilities suggests that the mother-daughter relationship does not influence spatial behaviour under extensive conditions in these Spanish aggressive cattle.

Investigation of the vocalisation rate during the estrus cycle of dairy cattle

P.C. Schön[1], S. Dreschel[2], W. Kanitz[2] and G. Manteuffel[1], [1]Forschungsinstitut für die Biologie landwirtschaftlicher Nutztiere (FBN), FB Verhaltensphysiologie, Wilhelm-Stahl-Allee 2, 18196 Dummerstorf, Germany, [2]Forschungsinstitut für die Biologie landwirtschaftlicher Nutztiere (FBN), FB Fortpflanzungsbiologie, Wilhelm-Stahl-Allee 2, 18196 Dummerstorf, Germany

Based on results from other ungulates species and preceding own experiments revealing that there was an increase of vocalisation rate (number of distinct calls/hour or day) in tethered heifers during the estrus, we hypothesised that cows develop an altered vocalisation pattern during estrus. We recorded the vocalisation of 10 heifers (German Holstein, age of 14 to 26 months) while tethered and during free-range keeping in a pen with two other animals. Only sounds with a loudness above a pre-defined threshold were recorded and synchronised with time lapsed video recordings. In result were recordings of single time marked calls. We observed 31 natural estruses in tethered and 27 natural estruses in free-range. The results from tethering were similar as in the prior examinations. Nearly all animals (9/10) vocalised in 26/31 estruses with a peak in vocalisation rate on day 0 (=visual detection of estrus by human). The vocalisation rate on day 0 differed significantly from all other days (p<0.01, procedure GENMOD, repeated measures on animals, SAS) and there was an increase in rate from 12.2±3.8 (day -1) to 79.4±25.9 (day 0) and a decrease to 8.6±4.4 (day 1). However, beside a strong interindividual variability we also found a strong intraindividual variability in the animals (35.8±14.5–134.1±84.1 vocalisations on day 0) between the cycles. In free-range keeping all animals vocalised less than while tethered and not all animals (8/10 in 16/27 estruses) vocalised on the day of the estrus. On the average there was, however, an increase of vocalisation rate on day 0 (day -1: 11.4±6.6, day 0: 26.8±15.4, day 1: 3.2±2.3, p>0.1). Whether the lower vocalisation rate in free-ranging animals is due to other compensatory estrus behaviour will be a topic of future research.

How does male-female interactions affect puberty onset in heifers?

M.C. Fiol[1,2], G. Quintans[3] and R. Ungerfeld[2], [1]Facultad de Veterinaria, Departamento de Bovinos, Lasplaces 1550, Montevideo 11600, Uruguay, [2]Facultad de Veterinaria, Departamento de Fisiología, Lasplaces 1550, Montevideo 11600, Uruguay, [3]INIA Treinta y Tres, Ruta 8 km 282, Treinta y Tres, Uruguay

Little is known about the use of male exposure to advance puberty in heifers. Our first objective was to determine if heifers' body weight affect the response to androgenised steers. A second objective was to determine if physical proximity between heifers and steers –which is related to stimulating intensity- is associated to reproductive results. Sixty-six 12-month old Hereford x Aberdeen Angus prepuberal beef heifers, grazing on native pastures, were exposed to 4 androgenised steers during 35 days. Three categories were determined according to their initial body weight: low (< 211 kg, LW, n=22), medium (211 – 236 kg, MW, n=24) and high (> 236 kg, HW, n=20). Cyclic activity was determined through oestrous detection (40 min twice/day) and weekly ovarian ultrasound observations. Association index (AI) for each heifer was determined every 10 min during 4 h, 3 times/week. AI was considered as 1 when heifers were less than 1 body length from steers, 0.33 when were between 1 and 3 body lengths, and 0 when were separated more than 3 body lengths. Mean AI for each heifer was calculated from each day's mean until cyclic activity began. Frequencies were compared using χ^2 test, AI were compared with ANOVA after Bliss transformation. More HW than MW and LW heifers [15 (75%) vs 7 (29%) and 4 (18%) for HW, MW and LW, respectively, $P<0.01$] reached puberty. AI was greater ($P<0.05$) in HW (0.11±0.09) than MW (0.06±0.04) and LW (0.06±0.03). We conclude that the intensity of the stimulus (male-female proximity) and the percentage of heifers that reach puberty (began cyclic activity) was related with heifers' body weight when male exposure started. It remains to be determined if this closest relationship is determinant of the earlier response.

Influence of permanent versus restricted access to the calf on maternal behaviour, behaviour during milking and milk yield in dairy cows

E. Hillmann[1], R.A. Schneider[1], B.A. Roth[1] and K. Barth[2], [1]ETH Zurich, Institute of Animal Science, Physiology and Behaviour Group, Universitätsstrasse 2, 8092 Zurich, Switzerland, [2]Federal Research Institute for Rural Areas, Forestry and Fisheries, Institute of Organic Farming, Johann Heinrich von Thünen-Institute, Trenthorst 32, 23847 Westerau, Germany

Mother-bonded rearing of dairy calves provides permanent or restricted contact between cow and calf, and cows are milked additionally. The aim of this study was to examine the influence of different degrees of mother-bonded rearing on maternal behaviour, restlessness behaviour during milking and milk yield in dairy cows. During the first three months in milk, three treatment groups were tested: A) Permanent contact (n=11) - permanent cow-calf contact; B) Suckling twice daily (n=13) - 15 min cow-calf contact before milking; C) Control (n=24) - Cows and calves were permanently separated 1 day post partum. Maternal behaviour (proximity to calf/calf-area, licking/suckling the calf, searching calls) of cows in groups A and B was observed individually once weekly for four hours/day using 5min interval-sampling. In the parlour, behavioural indicators for stress reactions were recorded (kicking against milking equipment, vocalisation, defecation/urination) and a stress score was defined. Milk yield was registered during each milking. For statistical analyses, linear mixed effects-models were used. Cows from group B were more often in proximity to the calf-area in the beginning of lactation (13±2 vs 5±1% of scans, p=0.003) and showed more searching calls than cows from group A (20.2±8.4 vs 2.9±.6, p=0.006). In the parlour, probability for vocalisation or highest stress score was increased in cows from groups A and B (treatment×DIM p=0.033), especially in cows of group B. Milk yield was reduced significantly in groups A and B and decreased faster over the first three months of lactation than in group C (treatment×DIM p<0.001). Cows with twice daily calf contact searched intensely for their calves at the beginning of lactation, which can be interpreted at least as a short-term stress. However, mother-bonded rearing with permanent calf contact proved to be a welfare friendly management system from the point of view of cow behaviour.

Milk portion size and milk flow affects dairy calves' behaviour

P.P. Nielsen[1], M.B. Jensen[2] and L. Lidfors[3], [1]Swedish University of Agricultural Sciences, Dept. Animal Nutrition and Management, Kungsängens Research Centre, 753 23 Uppsala, Sweden, [2]University of Aarhus, Dept. of Animal Health, Welfare and Nutrition, P.O.Box 50, 8830 Tjele, Denmark, [3]Swedish University of Agricultural Sciences, Dept. of Animal Environment and Helath, P.O.Box 234, 532 23 Skara, Sweden

Cross-sucking (CS) by artificially fed dairy calves is a redirection of the calves' normal sucking behaviour and it occurs in close connection to milk meals. We examined how a combination of milk portion size (1 or 2 litres) and flow (300 or 600 ml per min) affects calves' use of a computer controlled milk feeder and the occurrence of CS. Forty-eight calves (43 Holstein-Friesian and 5 Swedish Red; 26 heifers and 22 bulls) were exposed to all four treatments in a cross-over design with four periods of 1wk each. The data were analysed by either a generalised mixed model or a variance component procedure. The calves' milk meals (rewarded visits) were longer ($p<0.001$) when they were fed the milk in large (8 to 10 min) compared to small portions (around 6 min). Treatment had no effect on the number of calves performing CS (20 out of 26 bull calves and 18 out of 22 heifer calves performed CS). More bull calves (26) received CS than heifer calves (17, $p<0.01$) and more CS was performed by heifer (1115 times) than bull calves (751 times, $p<0.001$). Subsequently, CS was categorised as occurring within 30 minutes after a milk feeder visit (i.e. related to the visit), or >60 minutes after (i.e. not related to a visit to the milk feeder). The CS occurring between 30 to 60 min after a milk feeder visit was removed from the analysis. Neither portion size nor milk flow affected the proportion of calves performing CS within the first 30 min after a visit. However, more calves performed CS >60 minutes after a feeder visit, when they were fed small milk portions ($p<0.05$) or when they were subjected to a slow milk flow ($p<0.05$). In order to allow calves to satisfy their sucking needs they should receive the milk in large portions to prolong rewarded visits and thereby reduce cross-sucking occurring more than 60 min after a visit to the milk feeder.

Thermoregulatory behaviour and feeding level of newborn dairy calves

F. Borderas[1,2], A.M. De Passille[3] and J. Rushen[3], [1]Universidad Autonoma Metropolitana-Xochimilco, Mexico City, Mexico City, Mexico, [2]University of British Columbia, Main Mall, Vancouver, BC, Canada, [3]Agriculture and Agri-Food Canada, PO 1000 6947 Highway 7, Agassiz, BC, V0M 1A0, Canada

Cold increases mortality of calves but we know little of calves' abilities to thermoregulate. To examine effects of feeding level on thermoregulatory behaviour, 17 Holstein calves were housed for 3d after birth in full floored, individual pens (3.8x2.0m) with wood shavings as bedding, and heat lamps at one end to provide a thermal gradient. They were fed milk at high (30% BW) or low (8% BW- a common commercial practice) daily allowances. Daily temperatures were 3.71±0.54 °C. Temperature loggers on each calf recorded ambient temperatures. Video cameras recorded the calf's distance from the heat lamp and lying posture (on sternum or side, with legs extended or contracted) for 24h/d. Calves spent more time (68% of time) in the one-third of the pen fitted with heat lamps than in the remainder of the pen (Proc Mixed $p<0.05$) but there was no effect of feeding level ($p>0.10$). Low-fed calves tended to be closer to the heat lamp during the coldest period of the day than high-fed calves (distance score=0.38 vs 0.53; $p<0.10$). For low-fed calves there was a correlation ($r=-0.39$ $p<0.001$) between the barn temperature and the calves' distance from the heat lamp, but there was no correlation for high-fed calves ($r=0.08$ $p<0.10$). The difference between the temperature recorded in the barn and the temperature recorded on the calf's back was negatively correlated with distance from the heat lamp ($r=-0.38$ $p=0.01$) showing that the temperature logger on the calf reflected the ambient temperature chosen. Time spent in different lying postures was not affected by feeding level or barn temperature ($p<0.10$). Young calves show a preference for warmer environments and this preference interacts with the amount of milk fed. Temperature recorders attached to the calves can measure calves' thermal preferences. Calves did not thermoregulate through changes in resting posture.

Natural weaning in beef cattle: an on-farm case study

F.C. Flower[1], D. Albertsen[2] and A.E. Malleau[1], [1]Animal Compassion Foundation, 550 Bowie St, Austin, TX, 78703, USA, [2]Valley Farm, Chitterne, Warminster, Wiltshire, BA12 0LT, United Kingdom

The practice of natural weaning in beef cattle production is often avoided due to concern over compromised health and fear that yearlings will re-suckle cows and compete with calves for milk. An on-farm study with 2 herds of Aberdeen Angus cows (n=56, 63) was conducted between March and December 2007. Cows and yearlings were kept together with calves born between April and June 2007. Health data and any behavioural observations of cows nursing an animal other than her own calf were recorded daily pre- and post-calving during daily herd inspections. To monitor and assess potential loss of fitness of mother cows over time body condition scores were recorded monthly. Incidence of calf disease was similar for study herds and a control herd without yearlings (pneumonia n=1, 1, 0; scours n=3, 4, 3, respectively). Almost 50% of cows in both herds were suckled at least once by an animal other than their own calf but this did not result in loss of cow condition post-calving. Mean ($\pm$ s.e.m.) cow condition improved post-calving in Herd 1 (2.7 vs. 3.0$\pm$0.1; t_{47}=5.69, P<0.001) and showed no change in Herd 2 (3.2 vs. 3.3$\pm$0.1; t_{39}=0.65; NS). Even cows observed nursing by another animal more than 5 times maintained a good level of condition after calving in both Herd 1 (2.6 vs. 3.0$\pm$0.1; t_{7}=5.87, P<0.001) and Herd 2 (3.0 vs. 3.2$\pm$0.1; t_{5}=1.34, NS). Pasture quality (mean DM $\pm$ s.d. for Herd 1 (314$\pm$46 g/kg) and Herd 2 (409$\pm$62 g/kg) and mean NDF $\pm$ s.d. for Herd 1 (687$\pm$75 g/kg) and Herd 2 (704$\pm$20 g/kg)) was consistent. Although a large proportion of yearlings were observed to suckle post-calving, cow condition and calf health were not compromised. This on-farm case study shows that natural weaning of beef cattle is a promising alternative to traditional weaning methods.

Similarity in reactions of veal calves to a novel object test and a human approach test

J.L. Rault[1], H. Leruste[2] and B.J. Lensink[2], [1]Purdue University, West Lafayette, IN 47907, USA, [2]Institut Superieur Agriculture (ISA) Lille, CASE, 48 boulevard vauban, 59046 Lille, France

Currently on-farm animal welfare monitoring systems assess both fear of novelty and fear of humans. However, performing similar natured tests is time consuming and might emphasise the same underlying biological origin. Our objective was to identify if a novel object test and a human approach test performed on veal calves generated similar outcomes. Fifty farms were observed: 35 with small group (284 pens total of five calves) and 15 with large group housing (74 pens total of 25 to 100 calves). At 13 weeks of age, calves' latencies to touch a novel object introduced into the pen were recorded. The same calves were tested 1½ hours later for their latencies to touch a same non-familiar person, standing at the pen's fence. Each test lasted 3 min. Data were analysed using a GLM model. No differences were found between small and large groups for the latency to touch the object (66.61 ± 2.34s vs. 74.48 ± 3.05s; P=0.11) or the latency to touch the human (59.16 ± 2.22s vs. 60.97 ± 3.05s; P=0.70). The latencies to touch the object and to touch the human were correlated at farm level (r=0.34; P=0.01) and pen level (r=0.16; P<0.01). In small groups, the latencies were positively correlated at farm level (r=0.32; P=0.06) and pen level (r=0.18; P<0.01). However, in large groups, the latencies tended to correlate at farm level (r=0.50; P=0.06) but not pen level (r=-0.04; P=0.75). Calves approached faster the human than the object (61.9 ± 2.28s vs. 69.97 ± 2.18s; P<0.01), probably due to an order effect of the tests. The latency to touch the object or the human did not depend on the housing system. Both tests were moderately correlated. These results suggest that the two tests reflect the same underlying biological phenomenon as a general fear response or curiosity. In the future, for practical reasons, only the human approach test which is easier to perform could be used to assess fear in a welfare monitoring system.

The relationship between prolactin and reactivity to acute stressor in heifer calves

K. Yayou[1], S. Ito[2] and H. Okamura[1], [1]National Institute of Agrobiological Sciences, Laboratory of Neurobiology, 2, Ikenodai, Tsukuba, Ibaraki, 305-8602, Japan, [2]Kyusyu Tokai University, Laboratory of Animal Behaviour, Kawayu, Minamiaso, Aso-gun, Kumamoto, 869-1404, Japan

Prolactin (PRL) has been reported to have anxiolytic action and inhibitory action on the reactivity of hypothalamic–pituitary–adrenal axis. The aim of the present study was to study the relations of basal PRL and PRL response to acute stressor with behavioural and cortisol responses to the stressor in heifer calves. At 2 weeks of age, 20 Holstein heifer calves were subjected to a 10 min open–field test (OFT) followed by a presentation of feeding bucket for 15 min in the open–field. If the calf contacted the bucket, a blast air was applied to its muzzle (surprise test: ST). Blood samples were collected via jugular catheter before and after the tests and plasma concentrations of PRL and cortisol were obtained. The relations of basal and %–increment of PRL with behavioural responses to each test and %–increment of cortisol were analysed using principal component (PC) analysis and Spearman rank correlations. We identified four PCs with eigenvalues greater than 2.0 that explained 55% of the total variation. The first PC, named general activity, had high positive loadings for locomotion and exploration during OFT or ST and %–increment of PRL, latency to contact the bucket during ST. The %–increment of PRL positively interrelated with locomotion during OFT or ST (OFT: rs=0.494, p=0.027, ST: rs=0.425, p=0.062). The third PC, named fearfulness, had high positive loadings for jumping during OFT and basal PRL, and high negative loadings for defecation, latency to contact the bucket during ST and %–increment of cortisol. These results suggest that, to the acute stressors, basal PRL might have a negative relationship with fearfulness and that PRL response might have a positive relationship with general activity in heifer calves.

Does the feeding behaviour of chickens change in relation to the length of a dark period?

L.R. Duve, B.L. Nielsen, S. Steenfeldt and K. Thodberg, The University of Aarhus, Faculty of Agricultural Sciences, Institute of Animal Health, Welfare and Nutrition, Blichers allé 20, P.O BOX 50, DK-8830 Tjele, Denmark

The activity level and feeding pattern of broilers are influenced by the prevailing light schedule. A period of darkness is needed to ensure sufficient rest and proper bone and eye development. Prolonged scotoperiods might, however, have welfare implications through an increased risk of hunger and footpad dermatitis. The aim of this study was to evaluate how the duration of scotoperiods affects feeding behaviour. Ross308 broilers (n=3216) were reared from day 1-37, and in equal numbers exposed to either a fixed light schedule (Treatment A; 16L:8D) or an intermittent schedule (Treatment B; 8L:4D). The feeding behaviour of undisturbed birds was video recorded through 24h once a week from week 3 (24 groups of 64 chickens), and measurements of contents from crop and ileum before and after a scotoperiod (n=192 chickens) were collected once a week from week 2. Finally, production parameters and the degree of footpad dermatitis were recorded. Data were analysed using a general linear mixed model (SAS). Results from week 3 are presented as an example. In the 3h prior to a scotoperiod significantly more chickens from treatment A than B were engaged in feeding behaviour (% ± SE: 6.8±0.4 vs 5.2±0.1; p=0.02), whereas immediately after the scotoperiod the chickens in treatment B had a higher feeding activity (% ± SE: 21.2±1.0 vs 18.0±0.6; p=0.004). The higher feed intake in treatment A prior to the scotoperiod was substantiated by a significantly higher content of the crop at that time (g ± SE: 30.9±2.7 vs 13.7±3.2; p<0.0001). The chickens on treatment B had an overall higher feed intake and weight gain, but no significant difference in footpad dermatitis was found between the treatments. In conclusion, the broilers modified their feeding behaviour according to the light schedule they were exposed to, and treatment A did not appear to exacerbate hunger or footpad dermatitis.

The effect of environment on spatial clustering in laying hens

L.M. Collins[1], L. Asher[1], D.U. Pfeiffer[1] and C.J. Nicol[2], [1]Royal Veterinary College, Veterinary Clinical Sciences, Hawkshead Lane, North Mymms, Hatfield, AL97TA, United Kingdom, [2]University of Bristol, School of Veterinary Science, Langford House, Langford, Bristol, BS40 5DU, United Kingdom

Being highly sociable, we may expect the environments in which we house laying hens to impact on group dynamics. Here, we looked at effect of pen environment (pen dimensions: length x width x height:1.15x1.0x1.92m) on spatial clustering. Twelve groups of four laying hens were tested under three environmental conditions – wire floor (W), shavings (Sh) and perches, shavings, nestbox and peat (PPN) (perch heights: (1)0.2m and (2)0.4m). Groups experienced each environment twice, for five weeks each time, in systematic order. Video recordings were made one day per week for 30 weeks. To determine level of clustering, we recorded positional data from a randomly-selected 20-minute excerpt per video. On screen, pens were divided into six equal areas (10 areas in PPN – one area per half perch). Every five seconds, the area that each bird was in was recorded. This method was used, as it provides information on individuals within clusters, area and local resources can be taken into account and it is a quick and easy method for data collection and application. Clustering score (CS) was calculated for each five-second recording as maximum number of birds in any one area divided by number of occupied squares. Expected CS in each area of W or Sh environments was 0.667 and in PPN was 0.48. Observed average CSs were greater than expected: 1.23 (W), 1.40 (Sh) and 1.26 (PPN) (Chi-sq: all $p<0.001$). Between the environments, there was an effect of perch and nestbox presence on CS (GLMM: $F_{1,237}=14.83$, $p<0.005$), with less clustering in PPN than W and Sh. There was also an effect of floor type (GLMM: $F_{1,255}=6.52$, $p=0.011$) - groups in Sh clustered most. Although birds clustered more than expected in all pen environments, the different resources provided in the pen environments were shown to have a significant effect on degree of clustering.

Effects of environmental enrichment in relation to the behaviour of turkeys

A.H.D. Letzguss and W. Bessei, University of Hohenheim, Stuttgart, Germany, Department of Farm Animal Ethology and Poultry Science, Garbenstrasse 17, 70593 Stuttgart, Germany

Lacking stimulation is assumed to cause cannibalism, aggression and low locomotor activity in growing turkeys that are kept under commercial conditions. Therefore it can be expected that environmental enrichment stimulates locomotor activity and reduces damages caused by cannibalism and aggression. This study examines the effect of different enrichment elements on the behaviour of turkeys. Two open-sided houses with 5,541 (House A) and 6,476 (House B) turkeys (B.U.T. Big 6) were enriched with raised platforms, round and square bales of straw and wire baskets filled with hay. One separate turkey house with 4,236 male turkeys was left as unenriched control (House C). The animals were observed four times by scan sampling at four weeks intervals starting from six weeks of age. In addition enriched and unenriched areas were video recorded at the same observation periods. The videos were analysed by continuous observation (30 min) of 10 birds at a time in each of the distinguished areas. Differences in the behaviour of birds in the enriched and unenriched areas were tested using Wilcoxon -Test for nonparametric traits (JMP, 5.01 SAS Institute). The enrichment structures influenced the resting behaviour. In both enriched houses the total time of locomotor activity was significantly lower on square bales of straw ($p<0.0001$ pen A; <0.001 pen B) and on raised platforms ($p<0.0001$ pen A, <0.01 pen B) as compared to the free space. There was, however, no significant difference in the total locomotor activity between the enriched and unenriched houses ($p=0.1747$). Aggression and cannibalism were extremely low in all houses. The enrichments had no positive effect on the locomotor activity. The turkeys prefered the raised platforms and square bales of straw for their rest periods.

Visual human contact during rearing and adulthood reduces fear of humans in caged laying hens

L.E. Edwards[1], P.H. Hemsworth[1], G.J. Coleman[2] and N.A. Botheras[3], [1]Animal Welfare Science Centre, Department of Agriculture and Food Systems, University of Melbourne, Parkville, Victoria, 3010, Australia, [2]Animal Welfare Science Centre, Department of Psychology, Monash University, Clayton, Victoria, 3800, Australia, [3]Department of Animal Sciences, The Ohio State University, 222E Animal Science Building, 2029 Fyffe Road, Columbus, Ohio, 43210, USA

Poor handling may limit farm animal productivity and welfare. This experiment examined the effects of three handling factors in a factorial design on 287 laying hens: (1) routine or an additional 12 minutes daily of visual contact with a human during rearing; (2) 30 s daily contact with a stationary human or 5 s daily contact with a human moving quickly and unpredictably from 18 weeks of age; and (3) this human contact from 18 – 38 weeks of age was at a distance of 0, 60 and 120 cm from the birds. From 18 weeks, the following measurements were taken: avoidance of an approaching human by observing presence or absence of bird at the cagefront during four standard approaches by the experimenter; bodyweight; and plasma corticosterone response to close visual human contact. Individual egg production records were also kept for each bird during this period. Logistic regression analysis showed that the proportion of birds at the cagefront during behavioural testing increased in response to additional contact during rearing (0.48 vs. 0.31, B=0.80, P=0.02), and adult human contact at a distance of 60 cm compared to 120 cm (0.48 vs. 0.34, B=-0.38, P=0.02). ANOVA showed that birds receiving human contact at 60 cm also weighed more than those at 0 cm (1.35 kg vs. 1.32 kg, F=3.21, P=0.04). Egg production was higher for birds exposed to the stationary human from 18 weeks (91.4% vs. 90.2%, F=5.47, P=0.02), with a trend for these birds to have lower corticosterone concentrations (3.30 vs. 4.12 ng/ml, F=3.27, P=0.07). These results indicate that brief visual contact with humans during both rearing and adulthood can reduce fear responses, with possible implications for improved bird welfare and productivity.

Form but not frequency of beak use by hens is changed by housing system

T. Shimmura[1], T. Suzuki[1], T. Azuma[1], S. Hirahara[2], Y. Eguchi[1], K. Uetake[1] and T. Tanaka[1], [1]Azabu University, Veterinary Medicine, 1-17-71 Fuchinobe, Sagamihara, 229-8501, Japan, [2]Kanagawa Prefectual Livestock Industry Technical Center, Livestock Engineering, 3750 Hongo, Ebina, 243-0417, Japan

To verify the hypothesis that hens in different housing systems have the same time-budget when it comes to different beak-related behaviours, we compared the pecking behaviour in hens in six housing systems: small (SC, 2 hens/cage) and large (LC, 2 hens/cage) conventional cages, small (SF, 5 hens/cage) and large (LF, 18 hens/cage) furnished cages, single-tiered aviary (SA, 18 hens/cage), and free-range (FR, 18 hens/cage). The floor-reared 284 layers were randomly divided into six housing systems. The number of hens performing behaviours using the beak was recorded in each cage/pen at various ages up to 63 weeks of age. The experimental unit was group, and the data were analysed by using Friedman's test with replication followed by the Steel-Dwass' multiple comparison test. Grazing by a large proportion of hens was observed in FR (35.6%), and litter pecking was performed most frequently in SA (17.1%). Significant effects of housing type were found on the proportion of hens eating (χ^2_6=94.9, P<0.001), drinking (χ^2_6=68.9, P<0.001), and preening (χ^2_6=67.4, P<0.001), and these proportions were higher in SC, LC, SF, and LF (average 33.3%, 4.3%, 16.9%) than in SA (20.9%, 3.1%, 13.4%), and higher in SA than in FR (12.4%, 1.9%, 7.4%; all P<0.05). Housing type had a significant effect on the proportion of hens performing object pecking (χ^2_6=60.1, P<0.001), and the proportion was higher in SC and LC (average 6.5%) than in the other systems (average 3.2%; all P<0.05).The proportion of hens performing severe feather pecking was higher in LF and SA than in FR, and higher in FR than in SC, LC, and SF (all P<0.05). The proportions of hens performing all pecking behaviours were almost identical among the six housing systems (about 64.0%). We concluded that caged hens may express motivation for beak-related behaviour by directing it at food, nipple drinkers, their own feathers, and the cage wires, although feather pecking appeared not to be decreased simply by it.

Adrenal reactivity in lines of laying hens selected on feather pecking behaviour

J.B. Kjaer[1,2] and D. Guémené[3], [1]Friedrich Löffler Institute, Institute for Animal Welfare and Animal Husbandry, Dörnbergstrasse 25-27, D-29223 Celle, Germany, [2]University of Aarhus, Dept. of Genetics and Biotechnology, Blichers Allé, DK-8830 Tjele, Denmark, [3]INRA, Behavioural Biology and Adaptation in Poultry, Centre de Tours, F-37380 Nouzilly, France

Domestic chicken lines of the White Leghorn type differing in their level of feather pecking have been developed by divergent selection specifically on feather pecking behaviour. This paper describes an investigation of basal level (before 2 min after catching), reactivity to manual restraint (10 min after catching and restraining) and maximal adrenal response to 1-24 ACTH (10 min after injection) in plasma of breeder birds of the sixth generation of selection (S6) and their control line. Birds from the three lines (N=27 per line) had comparable basal levels of corticosterone (1.6 ng/ml, anova $F_{2,101}=0.62$, $P>0.05$), whereas males had higher levels than females, lsmean 1.9 vs. 1.5 ng/ml (anova $F_{1,103}=6.03$, $P<0.05$). Reactivity to handling and restraint for 10 minutes differed with HFP birds showing higher reactivity than LFP birds, lsmean 11.0 vs. 7.9 ng/ml ($t=-2.00$, $P<0.05$), while control birds showed intermediate levels (10.2 ng/ml). Males had higher reactivity than females, lsmean 11.2 vs. 8.2 ng/ml (anova $F_{1,103}=3.96$, $P<0.05$). Maximal reactivity did not differ between lines (average 35.7 ng/ml, anova $F_{2,101}=1.38$, $P>0.05$). Males had higher maximal reactivity than females, lsmean 41.3 vs. 33.6 ng/ml (anova $F_{1,103}=5.77$, $P<0.05$). The present study indicates that correlation between genes affecting the components of the HPA-axis responsible for adrenal response and genes affecting feather pecking behaviour in our experimental lines is low and there is good reason to believe this is a general effect in poultry. The direction of this correlation, if any, is so that selection against feather pecking behaviour will result in lower adrenal sensitivity to human handling and restraint. From an animal welfare point of view this is a positive relationship.

Effects of social rank and familiarity on dust-bathing of domestic fowl

T. Tanaka, T. Shimmura, T. Nakamura, T. Azuma, Y. Eguchi and K. Uetake, Azabu University, Animal Behaviour and Management, 1-17-71 Fuchinobe, Sagamihara-shi, 229-8501, Japan

The objective of the present study was to investigate the effects of social rank (Exp. 1) and familiarity of a conspecific (Exp. 2) on dust-bathing in laying hens. We offered birds a choice using a combination of stimuli and evaluated the quality and quantity of dust-bathing. Twenty-four medium-rank hens of 120 were selected as test subjects. The dust baths were placed at each side of the test cage, and the stimulus cages were on both sides to introduce the stimulus hens. A wire mesh was placed between the test cage and the stimulus cage. The stimuli presented were combinations of a high-ranked hen, a low-ranked hen, or no hen for Exp. 1, and a combination of a familiar hen, an unfamiliar hen, or no hen for Exp. 2. Twelve birds each were used as high-ranked, low-ranked, familiar and unfamiliar hens. The duration and number of dust-baths, wing tosses, and other behaviours were measured. The data were analysed by using Wilcoxon single-ranks test. For Exp. 1, the test hen performed dust-bathing more frequently ($P<0,05$) on the side of the hen regardless of its social rank when presented with a choice of a high- or low-ranked hen, or no hen, although no significant effect of social order was found. For Exp. 2, the test hen performed dust-bathing more frequently ($P<0,01$) on the side of the familiar hen when presented with a choice of a familiar hen, or an unfamiliar hen or no hen, and more frequently ($P<0,05$) on the side of no hen when presented with an unfamiliar hen and no hen. In conclusion, the presence of a high-ranked hen did not affect the frequency the dust-bathing. While dust-bathing was performed more frequently on the side of a familiar hen, the behaviour was restrained by the presence of an unfamiliar hen.

The relationship between comb and wattle lesions and body weight of laying hens in large or intermediate groups may give hints about social structure

A. Moesta[1], U. Knierim[1], A. Briese[2] and J. Hartung[2], [1]Faculty of Organic Agricultural Sciences, University of Kassel, Department of Farm Animal Behaviour and Husbandry, Nordbahnhofstr. 1a, 37213 Witzenhausen, Germany, [2]University of Veterinary Medicine Hannover, Institute of Animal Hygiene, Welfare and Behaviour of Farm Animals, Buenteweg 17p, 30559 Hannover, Germany

While in smaller groups of hens a social hierarchy based on individual recognition is expected, there are several theories regarding the social structure in large groups. One of them assumes a few "despotic" birds, and a large proportion of birds showing low agonistic behaviour. Following this hypothesis we expected despotic birds to have higher body weights due to effective fighting for resources. Taking comb lesions as an indicator of physical social interactions, we further expected that there would be fewer birds with several lesions in large groups due their higher number of tolerant birds. At 31 weeks of age we weighed and counted the number of birds with either no comb and wattle lesions or more than one minor lesion ('several lesions') from two pens of an aviary (each with 1250 hens and 12 cockerels, 2x50 birds sampled) and from four intermediate groups (furnished cages, group size 39-60, 195 birds sampled). Contrary to our hypothesis, 63% of the aviary hens had several lesions compared to 24% in cage groups. Correspondingly, 15% of the aviary hens had no lesions, while in the cages it was 55% (n=232, p<0.0001, χ^2=52.4, Chi-square test). Although a contribution of the cockerels in the aviary to the lesions during treading cannot be excluded, this result may be explained by individual recognition in the intermediate groups with less physical and more threat-avoidance interactions, and pecking of despotic birds in the aviary. However, aviary hens without lesions did not have significantly greater body weight (median, no lesions: 2020 g, several lesions: 1950 g, n=78, p=0.33, Z=0.97, Wilcoxon´s-two-sample test), thus not substantiating the hypothesis of despotic birds in the two large groups investigated. The suitability of comb and wattle lesions as an indicator of physical social interactions should be further investigated including counts on a continuous scale.

The relative motivation for full social contact in juvenile silver fox (*Vulpes vulpes*) vixens measured in an operant apparatus

A.L. Hovland[1], G.J. Mason[2], R.D. Kirkden[3] and M. Bakken[1], [1]Norwegian University of Life Sciences, Department of Animal and Aquacultural Sciences, P.O.Box 5003, N-1432 Aas, Norway, [2]University of Guelph, Department of Animal and Poultry Science, Guelph, Ontario, Canada N1G 2W1, Canada, [3]University of British Columbia, Faculty of Land and Food Systems, 2357 Main Mall, Vancouver, BC, Canada V6T 1Z4, Canada

Quantifying the strength of social motivation is important to determine whether silver foxes should be allowed social contact when commercially housed. Six 7-8 month old vixens were housed continuously in a closed economy operant apparatus and their 'maximum price paid' was measured when the animals worked to obtain unrestricted social contact or food. Six vixens of the same age and origin acted as stimulus animals during the social test. Both test and stimulus foxes could decide their own visit durations to a shared compartment. The vixens' motives for access to social contact were examined by recording their behaviour during visits to the shared compartment. When access was almost free the test vixens visited the stimulus fox 30.3 ± 4.3 times and spent 45% of the time together per 24h. The stimulus foxes visited the shared compartment 40.7 ± 9.6 times and spent on average 80% in this cage per 24h. When entrance costs increased, the test vixens were still willing to pay for access and their average maximum price paid for contact was $38.3\pm8.3\%$ of the price paid for food. During the time spent together, the stimulus foxes made use of their opportunity to leave the cage, on average 1.43 ± 0.20 times per visit. During initial encounters the vixens would fight to establish dominance. However, during subsequent interactions social behaviour was characterised by sniffing and grooming (23.7 ± 4.3 times/h), play signals (10.5 ± 4.1 times/h) and agonistic displays (5.5 ± 2.5 times/h). Not all of the time together was used for social interaction and during the observation period the vixens used $33.1\pm6.9\%$ of the time for synchronous resting. Our results show that young silver fox vixens were motivated for social contact, and that vixens' motives for contact was mainly non-aggressive and could benefit their welfare due the occurrence of grooming and play.

Behaviour of gilts and entire male pigs at the end of the finishing period in a restricted feeding system

S. Conte[1,2], L.A. Boyle[2], N.E. O'Connell[1] and P.G. Lawlor[2], [1]Agri-Food and Biosciences Institute, Hillsborough, Co. Down BT26DR, United Kingdom, [2]Teagasc, Moorepark, Pig Development Unit, Fermoy, Co.Cork, Ireland

Interest in rearing entire male pigs is growing in the EU due to the likelihood of a future ban on castration. However, entire males perform more aggressive and sexual behaviour which could lead to welfare problems, particularly where pigs are restrictively fed in the final finishing period prior to slaughter. The aim of this study was to evaluate the behaviour of restrictively fed gilts and entire males at the end of the finishing period (c. 80-100 kg). Five pens of gilts and 4 pens of boars (13 pigs/pen; stocking density: 0.73 m^2/pig) were observed at 50, 60 and 70 days (d50, d60, d70) after transfer to the finishing house. Pigs were fed a liquid diet 3 times per day. Each day behaviour (active, inactive, feeding) and posture (stand, lie, dog sit) of all pigs was recorded at five min intervals for 30 min prior to and for one hour after each feeding event. Log transformed data were analysed by mixed procedures for repeated measures ANOVA using SAS. There was no effect of sex on any of the variables ($P>0.05$). The proportion of time spent in active behaviour increased from d50 (0.51) to d70 (0.60) (s.e.: 0.06; t=-2.89; P=0.04). There was more standing prior to feeding on d70 (0.52) compared to d60 (0.49) or d50 (0.34) ($P<0.01$). The proportion of transitions between postures was higher before compared to after feeding ($F_{1,8}=20.64$; P=0.002). However, it was lower for both feeding stages on d70 (0.25) compared to d60 (0.28) (s.e.: 0.05; t=2.91; P=0.03). These results reflect higher levels of disturbance before feeding when pigs were heaviest. The unexpected lack of a quantitative difference in behaviour between gilts and entire males suggests that levels of disturbance were similar for both sexes at feeding. This could be explained by the somewhat generous space allowance provided.

Increasing gestation stall length does not alter sows' lying behaviour

J.M. Siegford[1], B.E. Straw[2], M.K. Sharra[1] and E.J. Distel[1], [1]Michigan State University, Animal Behavior and Welfare Group, Department of Animal Science, 1290 Anthony Hall, East Lansing, MI 48824, USA, [2]Michigan State University, Department of Large Animal Clinical Sciences, College of Veterinary Medicine, D202 Veterinary Medical Center, East Lansing, MI 48824, USA

Sow housing is a critical animal welfare issue facing the swine industry in the US. In particular, producers using gestation stalls must ensure that sows have good welfare. The National Pork Board PQA Plus program now requires that 90% of pigs must be able to lie in full lateral recumbency without their heads resting on a raised feeder. Gestation stalls too short for sows reduce welfare by forcing them to lie with heads on feeders. We hypothesised that a simple modification to existing gestation stalls to increase length by 30 cm would allow sows to lie more, and lie more in full-lateral recumbency. Twenty-four multiparous sows were evenly allocated by parity and body weight to control (CON) or expanded (EXP) gestation stalls after being confirmed pregnant. The rear gate of EXP stalls was opened to increase stall length by 30 cm/sow. Instantaneous samples of sow posture and behaviour were taken 1x/min for 2 hr each morning for two days before and after EXP opening and 2, 4, and 8 weeks later. Analysis with a mixed model procedure showed no treatment differences in subsequent reproductive measures (P>0.05 for all) between EXP and CON, including number piglets/litter, birth weight of litters, number of piglets weaned, or litter weaning weight. Increasing stall length (EXP) when compared with CON did not affect full- or semi-lateral lying (4.21±3.22 versus 3.61±3.15 and 18.06±4.18 versus 16.04±4.14 respectively), sternal recumbancy (10.64±2.46 versus 8.67±2.48), standing (65.63±8.40 versus 79.60±8.23) or dog-sitting (20.80±5.74 versus 12.15±5.60; mean counts ± SEM; P>0.05 for all). More lying of all types (48.02±6.35) and less standing (46.52±7.34) were observed at week 4 relative to earlier time points (P<0.01 for all). EXP sows could contact neighboring sows through the back gate of the stall, but tended to display less aggressive behaviour relative to CON sows (0.27±0.15 versus 0.65±0.16; P=0.1). Increasing stall length alone does not appear to change lying patterns of sows. Stalls may need to be widened to accommodate the deep bodies of modern sows.

The motivation of gestating sows for an enriched group pen and their behaviour after 24 hours of deprivation

M.R. Pittman[1], A.K. Johnson[2], J.P. Garner[1], R.D. Kirkden[1], B.T. Richert[1] and E.A. Pajor[1], [1]Purdue University, Animal Sciences, 125 S. Russel St, West Lafayette, IN 47907, USA, [2]Iowa State University, Animal Science, 2356G Kildee Hall, Ames, IA 50011, USA

Gestating sow motivation for enriched environments is unknown, but is essential knowledge for developing housing that promotes welfare. Social rank may affect motivation, with subordinate (S) sows being less motivated for an enriched group pen compared to dominant (D) sows due to negative social interactions (i.e. aggression). We hypothesised that D sows would show higher motivation for an enriched group pen and spend more time with enrichments compared to S sows. The aims of this study were to 1) compare the motivation of gestating sows (D, n=7; S n=7) for first access (23 hours) to a group pen with feeding stalls and 4 enrichments (a rubber mat [1.83m x 2.44m], cotton rope, compost, and straw) and 2) to evaluate the effects of social rank (D, Middle [M], and S) on sow behaviour. To measure motivation, sows were trained to press an operant panel on an increasing schedule. The highest schedule pressed was used as an indicator of motivational strength and did not differ for D and S sows (41±11.77 vs. 53±11.74, p>0.05). The behaviour of thirteen groups of 3 Landrace x Yorkshire sows (parity 1-2) was recorded following 24 hours of enrichment deprivation. Data were analysed using GLM and post-hoc Tukey tests. Following deprivation, sows spent 39.0±3.12% of time interacting with enrichments, which was similar to the testing session, 43.8±3.64%. Social rank affected enrichment use (GLM: $F_{2,24}$=3.50, p<0.05) and inactive behaviour (GLM: $F_{2,24}$=4.40, p<0.05); where D sows showed higher enrichment use than S sows (p<0.05), while S sows showed more inactivity than D sows (p<0.05). No differences in locomotion, aggression or other behaviours due to social rank were found (all comparisons, p>0.05). Although enrichment use differed between sows, social rank did not impact motivation, indicating that access to the enriched pen was of similar importance for dominant and subordinate sows.

Behaviour, plasma cortisol, and immune cell populations of pigs experimentally infected with *Salmonella typhimurum*

D.C. Lay Jr., M.H. Rostagno, S.D. Eicher, K.A. McMunn and J. Chaya, Livestock Behaviour Research Unit, USDA-ARS, 125 South Russell St., West Lafayette IN 47907, USA

This study was necessary to determine if sickness behaviour could be used to identify pigs carrying *Salmonella* versus those that quickly rid themselves of infection. Four-month old swine were either infected with *Salmonella typhimurum* (n=24) or served as non-infected controls (n=15). Behaviour and immune cell populations were measured because they have proven to be indicators of infection for other diseases. Response (lying, sitting, standing, latency to approach and touch) to a novel object were recorded using continuous, direct observation for a 2-min duration at 1000, 1400, 1800, and 2200 h on d -1, 0, 1 and 2. Scan samples were used prior to collection of blood samples to quantify lying and standing. Blood samples were collected via jugular veni-puncture at 0 h, 8 h, 16 h, 24 h, 48 h, then weekly until 6 wks post-infection. At 0800 on d 0, infected pigs were given 10^8 colony forming units/pig. Finally, pigs were euthanised to collect bacterial count data from lymph nodes. Data were analysed using mixed models with repeated measures and Wilcoxon tests when appropriate. Infected pigs spent more time lying sternal (34.1±3.1 vs 19.4±3.5%) and standing (33.18±3.8 vs 14.7±4.6%) than controls (P<0.01). Control pigs spent more time lying lateral than infected pigs (66.1±6.3; P<0.01). Latency to approach the novel object was shorter for infected pigs (42.6±2.9 vs 52.0±3.9 s; P<0.05). The number of monocytes tended to be lower in infected pigs (0.74±0.03 vs 0.76±0.03 x10^9 cells/L; P<0.08), with no differences for the number of neutrophils, leukocytes, lymphocytes, eosinophils or basophils (P>0.10) as compared to control pigs. These data indicate that the behaviour of the pig, as opposed to changes in cortisol or immune cell populations, is a more sensitive measure of infection with *Salmonella typhimurum*.

Selection for reduced aggression in pigs: what is the impact on activity and social behaviour three weeks after mixing, and on behaviour during handling?

R.B. D'Eath[1], S.P. Turner[1], R. Roehe[2] and A.B. Lawrence[1], [1]SAC, Animal Behaviour & Welfare, Sustainable Livestock Systems, West Mains Road, Edinburgh, EH9 3JG, United Kingdom, [2]SAC, Animal Breeding & Development, Sustainable Livestock Systems, West Mains Road, Edinburgh, EH9 3JG, United Kingdom

Aggression at mixing has negative impacts on pig welfare and production. We recently demonstrated that genetic selection for reduced aggression is feasible. Fighting and bullying post-mixing were moderately heritable, and skin lesion counts 24hrs after mixing could be used as a proxy trait (see Turner et al. abstract at this conference). However, selection to reduce aggressive behaviour could have wider impacts, if genetic correlations exist between traits. Such impacts may be negative in a practical (e.g. reduced handling ease) or ethical sense (e.g. non-aggressive pigs are unresponsive or inactive). To investigate these, we recorded other traits in the same population of 1660 pigs used by Turner et al. Subjects were 895 purebred Yorkshire pigs and 765 Yorkshire x Landrace of both sexes, housed in partially slatted pens with straw bedding. Three weeks after mixing, we made 24 hr observations of activity. Inactivity was weakly heritable ($h^2=0.06\pm0.02$) and negatively associated with bullying ($r_g=-0.28\pm0.17$), suggesting that under selection for reduced aggression, pigs would become slightly less active. Pigs' behaviour during handling was scored at three stages during weighing at 10 and 22 weeks. A greater diversity of scores and higher heritability ($h^2=0.29\pm0.02$) was found for entering the weigh crate than for behaviour in the crate ($h^2=0.13\pm0.01$) or on exit ($h^2=0.11\pm0.01$). Entering and leaving the crate easily had low positive genetic correlations with aggressive behaviours (fighting and bullying, r_g between 0.08 and 0.25), although aggressive pigs were also more active during weighing (r_g -0.23 to -0.33). Because of the low genetic correlations, selection for reduced aggression would have only a small negative impact on handling ease at weighing. In summary, selection for reduced aggression in pigs would slightly affect the other behaviours examined. Genetic correlations were quite low, indicating a low genetic response in these traits which could be controlled using a breeding index. Possible impacts on production traits (growth and reproduction) are being analysed.

The value of telemetry for measuring effects of snaring and saline injection on heart rate variability in pigs

A.M. Janczak[1], R.M. Marchant-Forde[2], J.N. Marchant-Forde[2], D. Hogan[3], D.L. Mathews[4], C. Dowell[4], L.J. Freeman[3] and D.C. Lay, Jr.[2], [1]Norwegian School of Veterinary Sciences, Dept. Prod. Anim. Clin. Sci., Oslo, N-0033, Norway, [2]USDA-ARS, Livestock Behavior Research Unit, West Lafayette, 47907 IN, USA, [3]Purdue University, Veterinary Clinical Sciences, West Lafayette, 47907, IN, USA, [4]Purdue University, Laboratory Animal Program, West Lafayette, 47907, IN, USA

The aims of this project were to evaluate electrocardiogram (ECG) quality recorded from twelve 4-month-old domestic pigs using intra-cardiac leads and telemetry and to determine how snaring and injection of saline influences heart rate variability (HRV). While under general anaesthesia, a bipolar, intracardiac lead was inserted into the right jugular vein and secured into the apex of the right ventricle. The post-operative analgesics used were buprenorphine and banamine. Animals were monitored until recovered. A telemetric device connected to the intracardiac lead was used to collect ECGs from the pigs before and after snaring, which was necessary for immobilising them, and intravenous injection of saline. Series of 512 successive interbeat intervals (IBI) were checked for error and subjected to time and frequency domain analysis of HRV using Art 4.0 and Excel software programs. ECGs were nearly free of noise and signal disappearance. Relative to baseline values prior to disturbance, snaring and saline increased heart rate (from 115±15 to 124±10; mean ± sd; Wilcoxon signed-ranks test score: $Z=19$; $p\leq0.03$), reduced total power (from 5796±4155 to 3056±2126; $Z=-22.5$; $p\leq0.02$), increased the ratio of Low Frequency to High Frequency power (from 0.13±0.14 to 0.32±0.46; $Z=19.5$; $p\leq0.05$) and reduced the square root of the mean squared differences of successive interbeat intervals (from 12±6 to 7±2; $Z=-24.5$; $p\leq0.01$), indicating that snaring and saline injection influenced heart rate variability, corresponding to increased sympathetic and reduced vagal activation. Telemetry monitoring thus proved to be effective in monitoring responses to a stressor and can be used to remotely monitor effects of stress in pigs.

Azaperone (Stresnil®) reduces anxious open field behaviour in pigs

R.D. Donald[1], K.M.D. Rutherford[1], S.D. Healy[2] and A.B. Lawrence[1], [1]The Scottish Agricultural College, Animal Behaviour and Welfare, Sustainable Livestock Systems, West Mains Road, Edinburgh, EH9 3JG, United Kingdom, [2]University of Edinburgh, Institute of Evolutionary Biology, School of Biological Sciences, West Mains Road, Edinburgh, EH9 3JT, United Kingdom

Azaperone (Stresnil®) is used to reduce fighting during mixing of unfamiliar pigs, to prevent savaging in sows, and to induce anaesthesia for veterinary treatment. Other behavioural effects have not been widely explored. The aim of this study was to investigate the use of azaperone as an anxiolytic for the validation of pig anxiety and fear tests. Since isolation is aversive to pigs; an open field test, which involves isolation, was used to induce anxiety in pigs. Twenty-four 6-week-old piglets were observed in an open field (provisioned with a ball and feeder) for 10 min, following treatment with either 1.0 mg/kg azaperone (A: n=12), or an equivalent volume of saline (S: n=12). Half from each treatment (n=6) were observed immediately after injection (t0), and half (n=6) were observed 20 min after injection (t20). Generalised Linear Mixed Models were used for analyses. Piglets given azaperone showed increased locomotion (A: 113.3; S: 82.33; Wald=6.05, P<0.014), spent more time with the ball (A: 14.75; S: 3.3; Wald=20.78, P<0.001) and feeder (A: 34.84; S: 3.96; Wald=13.90, P<0.001) and grunted less (A: 17.76; S: 57.8; Wald=16.38, P<0.001) than control piglets. An interaction was found between treatment and injection time for grunting (Wald=5.27, P=0.035), and time spent exploring the feeder (Wald=7.63, P=0.006). Azaperone-treated piglets observed at t0 grunted less, while azaperone-treated piglets observed at t20 spent more time with the feeder. There was no effect of gender. Our findings indicate that azaperone reduces anxiety-related behaviours, and may therefore be a useful tool for the validation of pig anxiety and fear tests.

Effect of intravenous general anaesthesia for castration on behaviour of male pigs

J. Baumgartner, B. Czech, C. Goessler and C. Leeb, University of Veterinary Medicine Vienna, Veterinaerplatz 1, A-1210 Vienna, Austria

Castration of male pigs without anaesthesia is the most common procedure across Europe to prevent boar taint in pork. This practice is questioned increasingly from an animal welfare point of view. The aim of this study was to evaluate the impact of intravenous general anaesthesia on the behaviour of male pigs at surgical castration. A mixture of Azaperon (2mg/kg) and Ketamin (25mg/kg) was administered by intravenous injection into the ear vein of manually restrained piglets (age: 6-33 d). Animals were castrated by two vertical incisions, antibiotic spray was used and animals were separated into recovery boxes. A total of 34 males pigs from 9 litters were observed during castration and during the first 60 minutes in the recovery box. The behaviour of these castrated males and their 58 female litter mates were analysed during the first 24 hours in the farrowing pen using direct observation with scan sampling (15 minutes interval). Data were analysed with descriptive statistics. Three out of four males showed pain-related vocalisation and twitching of legs in response to the injection, and 10% reacted with a strong defensive movement to the surgical castration. Sixty minutes after castration more than 50% of the pigs were found in a lying position and showed uncoordinated movements. During the first four hours in the farrowing pen castrated males were more active compared to the females (38% vs 23% of piglets moving). Two out of three castrates showed hanging and wagging tails for eight hours. The mortality after surgical castration with intravenous general anaesthesia was low (0.42%). It is concluded that intravenous general anaesthesia reduces acute pain at castration of male pigs. However the procedure itself creates pain and stress and does not solve the problem of long term post operative pain.

Playing-fighting-biting and weight gain around weaning in pig farms using group and individual sow housing

M. Špinka[1], J. Šilerová[1], R. Šárová[1], K. Slámová[1] and B. Algers[2], [1]Institute of Animal Science, Department of Ethology, Přátelství 815, 104 00 Prague - Uhříněves, Czech Republic, [2]Swedish Agricultural University, Department of Animal Environment and Health, Gråbrödragatan 19, SE-53223 Skara, Sweden

We compared play and aggression around weaning, and assessed their relation to piglet weight gain, on commercial pig farms using two housing systems: five group housing farms (GH, 6–11 lactating sows and their litters kept in a large straw-bedded pen, 8.6 m²/sow on average) and in five individual housing farms (IH, each sow and litter kept in a pen with less straw measuring 6.7 m²on average). Water was freely available and feed provisioning was similar, though not identical on the ten farms. Weaning was accomplished (without mixing litters) by removing sows at 5.5 weeks on average. Play, fighting (vigorous pushing accompanied by biting) and isolated biting were recorded in 16 piglets (2 males and 2 females per litter) from 4 litters (in GH farms belonging to the same group) in each farm on day before weaning (W-1), weaning day (W) and 5 days after weaning (W+5). All statistics were based on farm averages. There was no difference between GH and IH farms in frequency of playing, fighting or biting (GLM, $p>0.10$). On W-1 and W+5, playing correlated with fighting ($r_S=0.81$, $p<0.01$; and $r_S =0.65$, $p<0.05$, n=10 farms) but biting correlated with neither playing nor fighting. Farms in which piglets bit each other a lot on W-1 had a lower piglet weight gain between W and W+5 days ($r_S =-0.70$, $p<0.05$, n=10 farms). In farms with higher weight gain between W and W+5 days, piglets played and fought more on W+5 day ($r_S =0.78$, $p<0.01$; and $r_S =0.68$, $p<0.05$, n=10 farms). We conclude that (i) social piglet behaviour around weaning was not different between GH and IH farms; (ii) play and fighting (but not isolated biting) seemed to form one continuum; (iii) playing and fighting in weaned (nonmixed) piglets seemed to indicate good adaptation.

Drinking behaviour as a reflection of thirst, a stressor in transport of early weaned piglets

N.J. Lewis, S. Wamnes and R.J. Berry, University of Manitoba, Animal Science, 12 Dafoe Rd., R3T 2N2, Canada

Early weaned piglets (EWP), 16-21d, initially consume too little to meet physiological needs (eg. hematocrit: weaning - 39.0%; 2d - 41.4%; $P<0.01$). Consequently, one Code of Practice in Canada has allowed 24h transport, another, a more conservative, 12h. Previous studies measuring hematocrit (Control (C): 39.0%; 6h: 38.6%; 12h: 39.5%; 24h: 41.0; $P<0.01$) and drinking behaviour (d1 - C: 2.4%; 6h: 2.7%; 12h: 3.0%; 24h: 3.7%; $P<0.01$), did not show significant differences between 12h and 24h of transport. The objective of this study, as part of a larger study, was to use continuous observations to determine if a significant difference existed. All piglets (192), including C, were loaded, transported (<20 min) and unloaded; removing loading/unloading effects and reflecting commercial practice. After transport piglets were housed in groups of 4 in weanling pens. Drinking was assessed, using time lapse video, for 3 days post transport in h 0-2, 6-8, 12-14 and 18-20. Drinking was affected by transport on d 1 and other factors in the experiment, season and body weight, did not have interactive affects. Spilt-plot-in-time analysis (arcsine square route transformation) showed transported piglets spent more time drinking after transport (d1 - C: 0.5%; 6 h: 1.1%; 12 h: 2.0%; 24 h: 3.2%; $P<0.05$). Piglets increased drinking bout frequency after transportation for 6h (d1 - 26.3 vs day 2/3 19.9) and 12 h (d1 - 44.4 vs d 2/3 - 16.6; $P<0.02$). Piglets transported for 24h increased both frequency (d1: 55.2 vs d 2/3: 20.2; $P<0.01$) and bout length (d1: 17s vs day 2/3: 10s; $P<0.01$). This study showed duration of transport (12h vs 24h) produced qualitatively different drinking behaviour in piglets, suggesting a higher physiological stress placed on piglets transported for 24 h compared to 12 h.

Molar eruption and the onset of feeding behaviour in domestic piglets

A.L. Tucker and T.M. Widowski, University of Guelph, Animal & Poultry Science, Guelph, ON, N1G 2W1, Canada

Dentition has not traditionally been considered in any areas of commercial swine behaviour. However, studies on miniature breeds show that mastication of solid feed coincides with the eruption of molars. In commercial breeds, pre-weaning feed consumption has been shown to ameliorate the post-weaning growth check, but no consistent indicators of why piglets begin eating have been identified. The purpose of this study was to determine whether molar eruption is associated with the onset of feeding behaviour or feed ingestion in piglets 0-4 weeks old. At age 5 days, 233 nursing piglets (24 litters) were offered Cr_2O marked feed. Twice weekly, fecal swabs (indicating feed ingestion) and molar eruption were recorded. Weight was recorded weekly. Eruption had occurred when any portion of the tooth crown had emerged through the gingiva while time at the feeder was based on 6 hours of continuous video observations each day after dental exams. Repeated measures ANOVAs (SAS, Proc Mixed & Glimmix) were employed to test the effects of molar eruption, gender, and growth on duration of time at the feeder and feed ingestion. Mean eruption for maxillary molar3, mandibular molar4, and maxillary molar4 were 6.02±0.18, 7.22±0.22, and 24.45±0.39 days. Prior to 14 days, piglets with molars erupting spent less time at the feeder than those without molars ($F_{1,423}$=4.31, p=0.038 maxillary molar3; $F_{1,423}$=5.30, p=0.022 mandibular molar4). At 7 days, piglets having molars in opposing jaws touching (in occlusion), spent less time at the feeder ($F_{1,704}$=7.33, p=0.035), but by day 21 spent more time at the feeder ($F_{1,704}$=7.33, p=0.049). No relationship between molar eruption and feed ingestion were found. These findings suggest the oral stimulation of molars erupting may cause aversion to solid feed in piglets under 14 days of age. However, occlusion of molars at 21 days may attract piglets to solid feed. This is the first study to investigate the relationship between tooth development, feeding behaviour and feed ingestion, demonstrating the importance of including this aspect of developmental biology in feed related studies.

Social behaviour of weaned piglets exposed to enrichment objects

G. Martinez-Trejo[1], M.E. Ortega-Cerrilla[1], L.F. Rodarte[2], J.G. Herrera[1], F. Galindo[2], O. Sanchez[3] and R. Lopez[4], [1]Colegio de Postgraduados, Ganaderia, Carr. Mexico-Texcoco km 36.5, 56230 Montecillo, Estado de Mexico, Mexico, [2]Universidad Nacional Autonoma de México, Departamento de Etologia y Fauna Silvestre, Ciudad Universitaria, 04510, Mexico, D.F., Mexico, [3]Universidad Autonoma Chapingo, Sociedad Cooperativa Agropecuaria y Forestal Chapingo, Carr. Mexico-Texcoco km 38.5, 56230 Texcoco, Estado de Mexico, Mexico, [4]Universidad Nacional Autonoma de México, Departamento de Genetica y Bioestadistica, Ciudad Universitaria, 04510 Mexico, D.F., Mexico

Early weaning is a common practice in Mexican pig farms, this causes distress and fighting among piglets. The aim of this study was to evaluate if the presence of objects in the pen would reduce pen-mate directed behaviour in piglets weaned at 21 d of age. 576 piglets (6.20 ± 1.34 kg) were exposed to one of four enrichment treatments (8 replicates); (C) control: (P) plastic bottles (PVC); (R) ropes; (PR) bottles plus ropes. Behavioural observations and scan sampling with continuous recording of piglet activity at 10 minutes intervals were conducted during four consecutive months, in periods of eight days, from 14.00 to 2.00 h. The proportion of time walking, feeding, lying, grooming, biting bars, tail sucking, ears sucking, belly nosing, chewing ears and tail and fighting was recorded. Data were checked for normality and subjected to PROC MIXED analysis with repeated measurements. Piglets exposed to P (1.74), R (0.86) or PR (0.56) spent less time ($P<0.001$) bar biting than C (4.10). The same was observed for tail sucking (P 0.04; R 0.04; PR 0.02; C 0.54), ears sucking (P 0.24; R 0.11, PR 0.07; C 1.22), belly nosing (P 1.14; R 0.77; PR 0.58; C 2.93); chewing ears and tails (P 4.17; R 2.34; PR 0.78; C 28.13) and fighting (P 74.48, R 47.92; PR 35.42; C 195.57). Animals exposed to R dedicated less time walking (22.04) than those in P (27.52), PR (23.87) or C (27.81) and more time lying (P 72.48; R 77.96; PR 76.13; C72.19) whilst piglets in P (10.68) spent more time feeding than R (9.25), PR (9.06) or C (8.38), with no differences in grooming. Piglets exposed to enrichment showed less pen mate directed behaviour.

Examination of the behaviour of growing rabbits in different alternative rearing conditions

G. Jekkel and G. Milisits, Kaposvár University, Faculty of Animal Science, Guba Sándor u. 40., 7400 Kaposvár, Hungary

Three hundred and twelve growing rabbits were used in two repetitions to analyse their behaviour when reared in different sized pens (50 x 170 cm or 100 x 170 cm), different floor types (wire net or straw litter) and at various stocking densities (8, 12 or 16 rabbits/m²). This group was compared with the behaviour of rabbits kept in commercial cages (30 x 33 cm, wire net floor and 16 rabbits/m²). Twenty-four hour video recordings were made every week on the same day for the rabbits. The actual behavioural form of each rabbit was recorded in every 10 minutes. The effect of size of cage/pen, floor type and stocking density on the occurrence frequency of the examined behavioural froms was statistically analysed by the GLM method. Based on the results it was established that the occurrence frequency of the comfort and social behavioural forms was significantly ($P<0.01$) higher in the pens, as compared to that in the conventional cages (12.94% and 13.85% vs. 8.61%, and 6.40% and 5.41% vs. 3.92%, respectively). In spite of this, the stereotypic behaviour was significantly ($P<0.001$) more often observed in the cages (5.91%) than in the pens (0.01% and 0.31%). The other examined behavioural forms were not significantly influenced by the size of cages and pens. The type of floor had a significant effect also on the occurrence frequency of comfort and social behavioural forms. The frequency of the comfort behaviour was significantly higher (14.23% vs. 9.37%, $P<0.001$), while that of the social behaviour was significantly lower (3.90% vs. 6.58%, $P<0.001$) on the wire net floor, as compared to that on the straw litter. Stocking density had no significant effect on the behavioural forms examined. Regarding to animal welfare, it was concluded that the examined alternative rearing methods could be advantegous by decreasing the occurrence frequency of the stereotype behavioural forms.

Changes on the response to social separation in oestrus ewes

A. Hernández, A. Fierros, A. Terrazas, A. Olazabal, A. Medrano and R. Soto, Facultad de Estudios Superiores Cuautitlán, Universidad Nacional Autónoma de México, Posgrado, Carretera Cuautitlán Teoloyucan Km. 2.5 Cuautitlán Izcalli, 54714, Mexico

In sheep, a decrease in the social behaviour has been studied in detail only around parturition. It is possible, however, that during the oestrus ewes change their social preference in order to allow courtship and mount. Knowledge of this and other behaviours related may facilitate ewe oestrus detection by the man. This work was carried out to study changes in social behaviour in ewes displaying oestrus at the beginning of reproductive season (August). Forty multiparous dry Columbia ewes were allocated in either: 1) Control females showing no oestrus (n=20) or 2) Experimental females displaying oestrus (n=20). Oestrus was synchronised by intravaginal sponges (FGA, 30mg). Each ewe was placed into a 2m x 2m enclosure pen within the main flock farmyard. In the first part, ewe stayed 5 min surrounded by their conspecifics; in the second part, ewe stayed alone for 5 min; in the third part, ewe stayed another 5 min with a male. Vocalisations, nosing, locomotion, eliminations and jumps were recorder. An agitation index was made considering all these variables; response to social separation was constituted by differences in each part of the test. Nosing to conspecifics was different between with-conspecific and with-a-male in the experimental group (6.5 ±1.3 vs 3.4 ±0.5, $P<0.001$). In the same parts of the test control ewes differed in eliminations to experimental ewes (0.1 ±0.0 vs. 0.2 ±0.1; 1.8 ±0.3 vs. 0.8 ±0.2, respectively, $P< 0.04$). Also, nosing to objects was different between groups ($P<0.001$) in the first and second part of the test. Finally, response to social separation was different between groups in the third part of the test: 9.4 ±0.6 vs. 7.7 ±0.4 ($P<0.05$). In conclusion, ewes in oestrus display a less intensive response to the social separation test in the presence of the male.

How can pre- and post-ingestive stimuli affect intake and feeding behaviour in sheep fed hay diet ?

A. Favreau, R. Baumont and C. Ginane, INRA, UR1213 Herbivores, Site de Theix, 63122 Saint-Genès-Champanelle, France, Metropolitan

The learning theory of diet selection in ruminants gives the major role to post-ingestive consequences, giving to pre-ingestive cues only a discrimination role. Greenhalgh and Reid (1971) highlighted the importance of pre-ingestive informations on sheep foraging behaviour by feeding the animals partially through a rumen fistula. One food was consumed orally ad libitum (o) and the same or another one was introduced intra-ruminally (r) in an amount equal to half the total quantity of feed taken on the previous day. Using the same method, we investigated the relative importance of pre- and post-ingestive stimuli by measuring various behavioural and digestive parameters. Our experiment involved six rumen fistulated sheep and two foods: Lucerne (L) and mixed grass (G) hays. The four treatments (Go/Gr, Go/Lr, Lo/Gr, Lo/Lr) were tested according to a Latin square design. Daily dry matter intakes (DMI) significantly increased between treatments (SAS ProcMixed, $p<0.01$) from 596 (Go/Gr), 727 (Go/Lr), 795 (Lo/Gr) to 919 g (Lo/Lr). Diet fill effect decreased from Go/Gr to Go/Lr and from Lo/Gr to Lo/Lr because L had a lower fibre content than G (480 vs. 630 g/kg DM). However, DMI increase seems to result from different behavioural adjustments: when G was orally consumed, sheep tended to eat longer (161 vs. 187 min, $p=0.08$) but not faster; in contrast, when L was orally consumed, sheep did not eat longer but tended to eat faster even if not significant (5.3 vs. 6.1 g/min, $p>0.1$). Thus, behavioural responses to this change in post-ingestive fill effect depended on the food consumed. Furthermore, at a similar post-ingestive diet fill effect (Lo/Gr vs. Go/Lr), the difference in voluntary intake should also be related to pre-ingestive cues: a higher motivation to eat L due to prior learning of nutritional characteristics and/or an easier mastication of L that allows higher intake rate.

Animal behaviour as an indicator of forage abundance

D.T. Thomas, M.G. Wilmot and D.G. Masters, CSIRO, Livestock Industries, Private Bag 5, 6913, Australia

Spatial distribution of sheep has been related to feed supply in rangeland grazing systems. However, the use of behavioural indices to predict the quality and quantity of feed in agricultural production systems has received little attention. We hypothesised that the behaviour of sheep grazing a standing wheat crop residue would change predictably as the available edible dry matter was depleted. For four weeks, 364 five year old Merino ewes grazed a wheat crop residue in a 90 ha paddock in Western Australia (31°35'S, 117°28'E). Four of the ewes were selected randomly and fitted with WildTrax GPS collars (Bluesky Telemetry Ltd, Scotland). The collars were programmed to record GPS location, temperature and roll and pitch angle (vertical and horizontal head movement sensors) at five minute intervals over the duration of the study. Twenty x $0.1m^2$ pasture quadrats were collected weekly along a W-transect and pasture components were oven-dried at 65 °C for 24 hours, sorted into dry forage, green forage and wheat grain, then weighed. Relationships between forage biomass and distance travelled, mean distance between sheep with collars, collar roll and pitch angle and mean daily temperature (MDT) were tested using multiple linear regression analysis (Genstat version 10). Grazing activity in sheep decreased over time as the amount of forage available in the pasture was depleted by grazing. Dry forage biomass (Y kg/ha) was positively related to distance travelled (X_1 km/day) and roll angle (X_2 vertical rotation arc°), with MDT (X_3 °C) as a covariate (Fitted regression model; Y=- 115 + 43.5X_1 + 20.8X_2 – 30.2X_3; $F_{3,68}$=36.5; R^2=60.0, P<0.001; X_1, X_2 and X_3 t pr. <0.001). Variance in dry forage biomass that could be explained by animal behaviour was higher for behaviour measured during the morning compared to the afternoon (54.5 v 22.6%). This study demonstrates that behavioural indices that are readily measured using GPS collars provide useful information about the animal's environment. Production, welfare and environmental outcomes could be improved by using animal behaviour as a real-time biological monitor.

The evaluation of a risk assessment approach to quantify animal welfare in response to flystrike, mulesing and alternatives to mulesing in sheep

A. Giraudo[1,2], N.W. Paton[3], P.A.J. Martin[3] and A.D. Fisher[1], [1]CSIRO Livestock Industries, Locked Bag 1, Armidale NSW 2350, Australia, [2]Agroparistech, 16 rue Claude Bernard, 75006 Paris, France, [3]Department of Agriculture and Food, Western Australia, 3 Baron-Hay Court, Perth, WA 6151, Australia

In Australia, flystrike can severely compromise sheep welfare. Traditionally, the surgical practice of mulesing was performed in lambs to alter wool distribution and skin conformation of the breech and reduce flystrike risk. The aim of this study was to use published behavioural and physiological data to evaluate the effectiveness of an epidemiologically-based risk assessment model in comparing lifetime sheep welfare for mulesing, mulesing with post-operative pain relief, and no mulesing (with and without significantly increased anti-fly chemical use). We used 4 measures: abnormal behaviour, cortisol, haptoglobin and bodyweight change. Using published data from mulesing as an example, lambs spend 20% of the time in hunched standing and cortisol reaches 135 nmol/L in the period afterwards. The incidence of flystrike is 5% for mulesed animals in the high rainfall zone (compared with 60-70% for unmulesed sheep in the absence of preventative chemical insecticide treatment). All data were linearly normalised to a common cardinal scale ("welfare impact"), from 1-10, based on the range between the highest and lowest responses for each variable. Severity of welfare challenge was calculated as $S=\Sigma I\times P$, where I=welfare impact and P=probability of that impact (e.g. flystrike risk). Over 5 years of life, the highest severity of welfare challenge was for unmulesed animals (21.9), followed by mulesing (11.8), mulesing with pain relief (9.3), with the lowest welfare challenge for no mulesing combined with increased chemical use (3.3). The risk assessment model yielded data that matched existing practical knowledge, in that when initially developed, mulesing, despite being a severe acute welfare challenge, represented a lifetime welfare gain for the sheep in comparison with no mulesing. With conventional mulesing scheduled to cease in 2010, any alternative that is more welfare-friendly in the short term will also have to be very effective against flystrike throughout life, to optimise animal welfare.

Can we detect stress-induced cognitive bias in farm animals? Development of a method for sheep

R.E. Doyle[1,2], A.D. Fisher[2], G.N. Hinch[1], A. Boissy[3] and C. Lee[2], [1]University of New England, University of New England, Armidale, NSW 2351, Australia, [2]CSIRO Livestock Industries, Locked Bag 1, Armidale NSW 2350, Australia, [3]INRA, Unit de Recherche sur les Herbivores, St Genès-Champanelle, 63122, France

An individual's cognitive evaluation may be altered according to its mental state. By providing an ambiguous situation for an animal to assess, insight into their mental state may be able to be obtained. In order to test for this effect in sheep, twenty ewes were trained to learn that a feed bucket placed in one pen corner resulted in a positive reward on approach; however, when the bucket was in the alternate corner, a negative reinforcer (sight of a dog) was presented. Once reaching the learning criterion, sheep were assigned to treatment or control groups. Treatment involved a restraint and isolation stressor (RIS) for 6h/d on three consecutive days. Putative stress-induced cognitive bias was tested by providing ambiguous "probe" buckets in three positions between the positive and negative buckets. Sheep approach behaviour to the five bucket positions was recorded. Animals were tested before RIS, following daily RIS, and the day afterwards. Plasma cortisol and behavioural responses in a novel arena were also measured. A graduated response in approach behaviour to the buckets indicated that the testing scenario successfully made the animals "assess" the probe situations. Split-plot ANOVA revealed that cortisol concentrations were elevated in RIS sheep compared with Controls (134.1 vs. 19.2 nmol/L; $F_{5,270}=31.90$; s.e.=7.14; P<0.01). There was a tendency for RIS sheep to have fewer bleats per 3 min (2.13) than Controls (3.5) in the arena test ($F_{1,17}=3.44$; P=0.08; s.e.=0.52). Despite these effects, there were no treatment differences following Chi-square analysis in whether sheep approached the various bucket positions ($X^2<0.5$; df=1; P>0.5). While the model successfully induced a physiological stress response in the sheep, it did not alter individual cognitive processes as measured by our test. Alternatively, due to their evolution as a prey species, sheep may be resistant to exhibiting behavioural changes in response to chronic stressors.

High-energy food supplementation in the last two weeks of pregnancy improves the viability of newborn goat kids from malnourished mothers

R. Santiago[1], H. Sánchez[2], N. Serafin[3], R. Soto[2], A. Olazabal[2], A. Medrano[2] and A. Terrazas[2], [1]Universidad Autónoma del Estado de México, Carrera de Medicina Veterinaria y Zootecnia Unidad Zumpango, Zumpango, Estado de México, 54700, Mexico, [2]Universidad Nacional Autónoma de México, Secretaria de Posgrado, Departamento de Ciencias Pecuarias, FESC, kM 2.5 Carretera Cuautitlán-Teoloyucan, San Sebastian Xhala, Cuautitlán Izcalli, 54714, Mexico, [3]Universidad Nacional Autónoma de México, Departamento de Neurobiología Conductual y Cognitiva, Instituto de Neurobiología, Km 15 Juriquilla, Querétaro, Qro., 56000, Mexico

We investigated whether viability of newborn kids from malnourished goats may be improved by a maternal high-energetic supplementation two weeks before parturition. Multiparous mixed-breed dairy goats were allocated in the next groups: 1) Control (C, n=11); 2) Malnourished during second half of pregnancy with only 70% or requirements (MN n=12); 3) Malnourished but supplemented, two weeks before parturition, with maize (S, n=14). 1) dystocia, 2) motor activity and head reflex response in the first hour after birth, 3) birth weight and 4) percentage of mortality during the fist 45 days after birth were recorded. Dystocias was significantly higher in MN compared to S and C groups (MN: 16/29, S: 7/32 and C: 3/23 kids, P<0.01). After birth, kids from MN goats spent longer time trying to stand up than kids from C mothers (1380.2±291.2 vs 464±75.9 sec., P=0.009), while non significant differences were found between kids from S and MN, and those from S and C goats. Similar results were found in the kid´s latency to be completely stood up (2457.4±243 vs. 1461.2±199.4 sec., P=0.01). Proportion of kids showing head rising after a touch on their nose was smaller in MN than in C and S (MN: 7/16, C: 14/20 and S 20/27, P=0.04). Kids from C were heavier at birth than those from S and MN mothers (C: 3.54±0.1, S: 3.04±0.9 and MN: 3.02±0.1 kg, P=0.01). Mortality from birth until the first 45 days of life was significantly higher in kids from MN than those from C and S mothers (MN: 40%, C: 12% and S: 18%, P=0.05). It is concluded that a high-energetic food supplementation few days before parturition improves some aspects of vitality and consequently the viability of kids from underfed mothers.

Mid or late pregnancy shearing affects the vigor of the lamb and its chances of survival

G. Banchero and G. Quintans, National Institute of Agricultural Research, Meat and Wool Production, Ruta 50 Km 12 La Estanzuela, Colonia, 70000, Uruguay

Shearing ewes during mid- or late pregnancy is associated with increased lamb survival under pastoral conditions. One reason for this seems to be an increased birth weight of the lamb. However, the magnitude of the response in birth weight has been variable across studies suggesting that this is not the only mechanism involved. We tested the hypothesis that ewes shorn at 70 days (mid) or 120 days of pregnancy (late) would give birth to more vigorous lambs–irrespective of their birth weight- compared to unshorn ewes. Fifty-seven Corriedale ewes bearing twins were allocated to three treatments: shearing at 70 days of gestation (P70, n=18), shearing at 120 days of gestation (P120, n=20) or unshorn (US, n=19). Ewes grazed native pasture until housed and fed individually to requirements during the last 10 days of pregnancy. Observations on lamb behaviour were recorded during their first hour of life. Then, lambs were bled and weighed. Lamb birth weight and blood glucose were compared using ANOVA. Frequency of lambs standing or suckling was adjusted by birth weight and compared using χ^2 test. P70 lambs were heavier than P120 which were heavier than US lambs (3.5, 3.4 and 3.1 kg; (se=0.1) P<0.0001). Thirty-five percent of P70 and 36% of P120 lambs stood in their first hour of life compared to 25% of US lambs (P<0.05). Twenty-four percent of P70 and P120 lambs suckled compared to 8% of US lambs (P<0.001). All lambs had similar blood glucose concentration one hour after birth (1.8, 1.7 and 1.9 mmol/L (se=0.12) for P70, P120 and US lambs). Prepartum shearing not only increased the birth weight but also had a significant effect on the lambs' behavioural activity.

Vocal structure between ewes and lambs during a period of three weeks after birth
H. Naekawa, Senior High School Attached To Shonan Institute of Technology, Tsujido Nishikaigan 1-1-25, 251-8511 Fujisawa, Japan

The objective of this study was to determine the day–to–day changes in the sound of the five pairs of ewe and lamb no older than three weeks after birth. They were transcribed into syllabic scripts as represented in International Phonetic Alphabet in order to examine the phonetic structure. After repeated listening to the playbacks as well as studying the vocal waveforms and the sound spectrogram, the calls were divided into three types of bleating for each group /ŋŋŋŋ/, /ɲaee/, /ŋnaeee/ for ewes and /ŋŋŋŋ/, /ɲeee/, /ŋneeee/ for lambs. Sample size were recorded in each number /ŋŋŋŋ/ (N=11), /ɲaee/ (N=212), /ŋnaeee/ (N=112) for ewes and /ŋŋŋŋ/ (N=9), /ɲeee/ (N=23), /ŋneeee/ (N=65) for lambs The items subject to analysis included duration of each vocalisation (vocalisation over time), fundamental frequency (pitch of the sound), sound pressure (intensity) and formant (a component of frequencies representing resonating sounds in the vocal tract). The vocalisation was gathered and recorded in a small flock by capturing the vocalisation scenes and behaviours with a video camera. The analogue to digital information analysis of the vocalisation of the sample sheep was performed as follows: the sounds were sampled based on the observation records of vocalisation and behaviour. The segments of vocalisation were input in a laptop computer via the USB audio interface, and then analysed by using the sound processing software (Sugi Speech Analyser). In ewes, the duration (0.5 sec) of the /ŋŋŋŋ/ type were shorter than the other types, the fundamental frequency (196 Hz), sound pressure (33 db), and first formant (1137Hz) of the /ŋŋŋŋ/ type were higher than the other types (P<0.05). A significant correlation was observed between the fundamental frequency and the first formant for the /ɲeee/ and /ŋneeee/ type calls among lambs (P<0.05).

Male-male mount in prepubertal lambs: relation with social rank

R. Ungerfeld, L. Besozzi, J. De León and S.P. González-Pensado, Facultad de Veterinaria, Departamento de Fisiología, Lasplaces 1550, Montevideo 11600, Uruguay

In previous experiments we observed an intensive male-male sexual activity in 5-6 mo old domestic lambs, positively related with the display of sexual behaviour towards oestrous ewes. Other authors, observing groups of lambs with different body size and ages, suggested that male-male mounts are related with social rank. The objective of this experiment was to determine if male-male sexual behaviour was related to social rank in a group of similar aged male lambs. Male-male mounts were weekly recorded (week 4-40) during 8 h in a single group of 13 male lambs born in the same week. Both, mounting and mounted lambs were identified. Individual social rank success indexes were determined using the food competition test when lambs were 3 and 6 months old. A total of 294 male-male mounts were registered. Each lamb mounted on 22.6±5.7 (range 0 to 68) occasions, and was mounted 22.6±6.2 (range 2 to 79) times. Seventy-two different interactions were recorded. Total number of male-male mounts in which each lamb took place was positively related with the success index (r=0.48; P=0.05). High ranked lambs (HR; five with greater success indexes) took part in more mount interactions than the five lambs with lesser success indexes (LR): 57.6±16.0 vs 21.2±6.8; P=0.03. While HR and LR mounted a similar number of times (27.0±11.2 vs 10.6±4.0), HR lambs were mounted more times than LR rams (30.6±6.6 vs 10.6±3.5; P=0.036). The significance of male-male mounting may be different in groups in which males differ in age/size than in those more homogeneous groups. However, in the present experiment high-ranked males were more active and participated more frequently in male-male mounts.

Characterisation of social play behaviour in female goat kids reared under natural and artificial conditions and its relation with the development and establishment of early social organisations

E.A. Lara[1], A.M. Sisto[2] and A.E. Ducoing[1], [1]FMVZ, UNAM, Producción AnimAl: Rumiantes, Ciudad Universitaria, Av. Universidad 3000, Coyoacán, 04510, D F, Mexico, [2]FMVZ, UNAM, Etología, Fauna Silvestre y Animales de Laboratorio, Ciudad Universitaria, Av. Universidad 3000, Coyoacán, 04510, D F, Mexico

The aims of this research were to: 1) Characterise social play and agonistic encounters during the nursing stage, 2) Characterise social organisation strategies, during 15 days post weaning. In Phase 1, two treatments were assigned randomly to 4 day-old female goat kids (n=41): artificial rearing (AR) and natural rearing (NR), with three repetitions each. Behavioural samplings (36h) of social play and agonistic encounters were made. Proportions of time per period of observation were compared. Results showed, that despite a tendency in AR kids to spent more time playing socially (P=0.06), they displayed more agonistic encounters (P<0.001). In Phase 2, three treatments were assigned randomly to female goat kids that came from AR (n=13), NR (n=14) and a mixed group (AR/NR) (n=14), with two repetitions each. Behavioural samplings (93h) of social play and agonistic encounters were made. Proportions of time spent playing socially were compared again, and agonistic encounters with their consequences were analysed. Success Index was obtained individually. Index of linearity for the dominance hierarchy was determined and graphed. There were no differences in proportions of time spent playing socially between treatments (P>0.05). Hierarchically high-ranked AR kids showed less agonistic encounters (P=0.03), and were the treatment with less displacements with contact (P=0.005). Hierarchically high-ranked NR kids displayed more interactions where they started but they did not displace (P=0.001) and were the treatment with more individual aggressions (P=0.004). The AR treatment showed suggestive values of imperfect linear hierarchy (h=0.82 y h=0.77). We conclude that learning from agonistic encounters during the nursing stage allow kids to form nonlinear dominance hierarchies, with less aggression in their agonistic encounters during weaning.

The social life of pet cats outside the home environment

C.C. Alberts and F.A. Pereira, Unesp - São Paulo State University - Campus of Assis, Ciências Biológicas (Biology), Av. Don Antonio, 2100, 19806-900, Assis, SP, Brazil

The domestic cat (Felis catus), is easily seen in large groups, either in urban or in farm environment, and research work by recent authors proved that cat sociality is quite complex. Studying domestic cat's social behaviour can provide insights on the relationship between these pets and humans and also can help the understanding about the factors that favours sociality in the Felidae. This work aimed to understand the social relations among pet cats when outside their residences. Pet cat, here, describes a cat that lives in an urban environment with shelter and food provided by its owner. The behaviour of adult male (n=08) and female (n=12) cats, with no defined breed was observed. Recorded behaviour included social categories of a focal individual and distances between focal and non focal subjects in a cat gathering site. The focal individual was randomly changed at the end of every session of 15 minutes. Each cat was observed for about 10 sessions, totalising about 750 hours, during eight months. No orthodox statistics methods were used and data were instead analysed using the Directed Tree sequential behaviour method, with the EthoSeq software. Social behaviour was evaluated using, among others, the Dominance Tree Method, with the Domina software. Both methodologies are based on Graph Theory and were used before only in studies on primate social behaviour. Sequential behaviour analysis and general social behaviour analysis indicate that affiliative social relations take place among cats of the same residence and/or among cats of different residences. Agonistic relations, however, were restricted to only a few individuals, generally among cats sharing the same home. Dominance analysis did not show linear hierarchy but some subsets of semi linear hierarchies. Results suggest that pet cats have intricate social interactions among each other when outside their owners' homes.

Short stay in a cattery does not produce separation anxiety related behaviour in cats

E. Creighton and J. Pilkington, Anthrozoology Unit, Biological Sciences, University of Chester, Chester, CH1 4BJ, United Kingdom

A large proportion of domestic cats are likely to spend short-term stays in a cattery. Cattery-related stress may predispose domestic cats to develop separation anxiety syndrome (SAS) on their return home, due to the cat's loss of confidence in their social relationship with their owner, particularly if owners fail to recognise separation anxiety related behaviour and attempt to punish unwanted behavioural changes. The aim of this study was to determine if short-term behavioural changes associated with feline SAS occurred in cats on their return home from a stay in a cattery. Frequency ratings of 21 SAS and other stress-related behaviours were surveyed from 48 owners one week before and one week after placing their cats for short stays (4 to 22 days) in commercial catteries across Northwest England. Wilcoxon Signed Ranks analyses revealed statistically significant increases in exploration behaviour and scratching appropriate objects ($P \leq 0.001$), and statistically significant decreases in play behaviour, cat-directed aggression and vomiting ($P \leq 0.001$). Rubbing behaviour showed an increase, but was not statistically significant after Bonferroni Correction ($P = 0.004$). All other behavioural items were stable within cats before and after the stay in the cattery. These behavioural changes are not consistent with clinical symptoms of separation anxiety in cats. The increase in exploration, scratching of appropriate objects (as distinct from destructive scratching), and the trend towards an increase in rubbing, suggest an increase in territorial behaviour that is consistent with the animal re-establishing its presence in the neighbourhood after its short absence. A decrease in play is associated with mild stress, though cat-directed aggression is expected to increase with stress. The decrease in vomiting may have been due to correct and regular feeding received at the cattery. There was no indication from this research that short stays in commercial catteries predispose cats to developing symptoms of feline separation anxiety.

Behavioural risk factors for aggression identified in 100 owned dogs that bit a person

E.N. O'Sullivan[1], A.J. Hanlon[2], B.R. Jones[2] and K. O'Sullivan[3], [1]Cork County Council, Veterinary, County Hall,, Cork,, Ireland, [2]School of Agriculture Food Science and Veterinary Medicine, Veterinary Medicine, Belfield, Dublin 4, Ireland, [3]University College Cork, Statistics, College Road,, Cork, Ireland

The behavioural and management history data provided by 100 owners of dogs that had bitten a person was analysed. The data were obtained by telephone interview of volunteer dog owners. Two sub-sets of data were created using 21 dogs having no history of perceived aggressive behaviour prior to the reported bite incident and 79 that had a record of aggression. Univariate logistic regressions were conducted to investigate the association between the likelihood of aggressive behaviour and owner or dog characteristics. Multivariate backwards stepwise logistic regression analysis was carried out using variables that had a P value of less than or equal to 0.25 identified through the univariate logistic analysis. Multicollinearity among independent variables in this model was investigated and found not to be a problem. The adjusted R^2 was 53.8%. All tests were two-tailed. Statistically significant behavioural predictors of aggressive behaviour included the dog initiating play (OR=3, 95%CI=1-9, P=0.02), not complying consistently with the command "sit" (OR=0.2, 95% CI=0.05-0.7, P=0.015), variable obedience to people (OR=4.21, 95% CI=1.54, P=0.005), variable obedience according to location (OR=3, 95% CI=1-8, P=0.07), displays of problem behaviour only when family members were present (OR=6, 95%CI=1-28, P=0.02), displays of fearful reactions in specific contexts (OR=5, 95% CI=1.6-15, P=0.004), displays of excessive or repeated behaviours (OR=4, 95%CI=1-16, P=0.028) and not being trusted in the presence of children (OR=6.1, 95% CI=1.9-20, P=0.003). On multivariate analysis, a significant association was identified between the presence of the behavioural variables lack of compliance with the command "sit",displays of fearful reactions in specific contexts and having a history of aggressive behaviour.

Aspects of the owner-dog relationship that may increase the likelihood of obesity in both partners

S. McCune[1], P.J. Morris[1], J.A. Montoya[2], I. Bautista[2], M.C. Juste[2], L. Suarez[2], C. Peña[2] and R.M. Hackett[1], [1]WALTHAM Centre for Pet Nutrition, Freeby Lane, Waltham-on-the-Wolds,, Melton Mowbray, LE14 4RT, United Kingdom, [2]Las Palmas de Gran Canaria University, Veterinary Medicine Service,, Faculty of Veterinary Medicine, Las Palmas, Spain

As in humans, obesity is a common nutritional disease in pets. Previous work demonstrated that owners of obese dogs were more likely to anthropomorphise their pets. The aim of this study was to identify aspects of the owner-dog relationship that may increase the likelihood of obesity in both partners. 122 dogs and owners were recruited and categorised into 4 weight status groups (A: overweight owner with overweight dog, B: lean owner with overweight dog, C: Overweight owner with lean dog, D: Lean owner with lean dog) according to owner BMI and canine body condition score. Psychological aspects of the relationship were also assessed by validated questionnaires. Owners of overweight dogs (groups A & B) were significantly more likely to use food to communicate rather than to meet energy requirements ($p<0.05$ Eta^2=0.728). They were also more likely to anthropomorphise their dog, by rating talking to it ($p<0.01$ Eta^2 =0.212) and having it in bed with them ($p<0.01$ Eta^2 =0.268) as more advantageous. This effect was stronger when the owners of the overweight dogs were overweight themselves (group A). Obese dogs' access to their owner's bed and higher levels of begging for food may indicate their ability to manipulate their owners and elicit food giving. These results agree with previous work in that owners of overweight dogs have a tendency to anthropomorphise their pets and show affection through feeding ($p<0.05$ Eta^2=0.728). For a number of questions the weight status of the owner exaggerated the effect. This may be due to overweight owners being more likely to have overweight pets. The interactions between obese parents and children, that lead to increased risk of obesity in children, may also be reflected in the owner-pet relationship. The owner-dog relationship is dynamic and both partners' behaviour needs to be considered.

The effect of time left alone at home on dog behaviour

T. Rehn and L.J. Keeling, Swedish University of Agricultural Sciences, Dept of Animal Environment and Health, Box 7038, 750 07 Uppsala, Sweden

Twelve privately owned dogs without any history of separation related behaviour problems were video-recorded at three different occasions when left alone in their home environment (T_1=0.5 hours; T_2=2 hours and T_3=4 hours) with the aim of investigating the effect of time left alone on dog behaviour. Video-recording started ten minutes before the owner left the house and continued until ten minutes after the owner returned, so that interactions between dog and owner as well as behaviour during separation could be studied. Data on heart rate (HR) were collected within the same time period in each treatment. Instantaneous recordings (15s-intervals) of behaviour were made in combination with continuous sampling of events. In addition to analysing behaviours separately, behaviours were also grouped together and defined as new variables; physically active included all non-stationary behaviours. If the dog was stationary but still alert or focused on something it was regarded as mentally active. Behaviour, compared between equivalent time intervals during separation, did not differ between treatments, although a number of differences were observed at reunion. Dogs showed a higher frequency of physical and mental activity (Kruskal Wallis:χ^2=7.31, d.f.=2, $P<0.05$ and χ^2=10.53, d.f.=2, $P<0.01$, respectively) in T_2 (0.37±0.07; 0.52±0.08, mean proportion of time ± SE) and T_3 (0.48±0.08; 0.48±0.07) compared to T_1 (0.20±0.07; 0.21±0.05). They also showed more greeting behaviour (χ^2=10.50, d.f.=2, $P<0.01$) and interacted more with their owners (χ^2=12.48, d.f.=2, $P<0.01$) in T_2 (0.27±0.08; 0.47±0.09) and T_3 (0.26±0.04; 0.42±0.09) compared to T_1 (0.09±0.04; 0.14±0.03). Mean heart rate was significantly higher (ANOVA: $F_{2,25}$=3.50, $P<0.05$) in T_3 (162±17.1, mean bpm ± SE) at the owner's return compared to the other treatments (T_1:112±6.7; T_2:120±11.3). According to the results of this study, the effect of time left alone, even within reasonable time frames, affected the responses of the dogs when the owner returned.

How well do staff and customers agree on quality of life indicators for dogs at re-homing centres?

D.B. Morton[1], L.M. Collins[2] and D.U. Pfeiffer[2], [1]University of Birmingham, School of Biosciences, Edgbaston, Birmingham, B152TT, United Kingdom, [2]Royal Veterinary College, Veterinary Clinical Sciences, Hawkshead Lane, North Mymms, Hatfield, AL97TA, United Kingdom

Quality of life (QoL) is a summation of poor and good welfare for an animal over time. Making a practical assessment of current and future expected QoL is of key importance in decision-making processes relating not only to euthanasia and care, but to decisions about housing design and living conditions for the animals in question. The aim of this study was to generate a list of perceived indicators of good and poor welfare that could potentially be easily and robustly recorded on a regular basis per animal to quantitatively assess its QoL. Staff at a dog re-homing centre (n=22) and members of the public visiting the centre (n=68) completed surveys providing information on level of dog-care experience (DCE), age, gender (staff) and group size (public), and listed the signs or behaviours they considered to be indicators of good and poor QoL in dogs. Age and DCE both affected types of responses given by staff (GLMM: Age: $F_{2,335}$=3.383, p=0.035; DCE: $F_{1,335}$=4.076, p=0.044; Gender: $F_{1,335}$=0.001, p=0.972). Age, DCE and group size had no overall effect on types of responses provided by visitors (GLMM: all p>0.05). However, visitors with more DCE suggested 'normal appetite' (t=3.15, p=0.008, df=12), and 'energetic' (t=2.992, p=0.007, df=12), more than average. Staff suggested more indicators of poor than good QoL (t=-4.775, p <0.0001, df=21). Visitors suggested more good than poor QoL indicators (t =2.458, p=0.017, df=67). Staff responses were ratio-scaled, summed with visitor response frequencies and ranked. A wagging tail was the top-ranking perceived indicator of good QoL. Behaviours considered 'stereotypic' were ranked collectively as the most commonly perceived indicator of poor QoL. It is likely that the responses were subject to memory-bias and audience effects on dog behaviour, potentially leading to a skewed representation of true QoL. Further work is needed to investigate these factors more fully.

Equine attraction to essential oil odours

J. Hurley and D. Goodwin, University of Southampton, Animal Behaviour Unit, School of Psychology, University Road, Highfield, SO17 1BJ, United Kingdom

There are a wide range of products containing essential oils. Aromatherapy for horses is becoming popular with owners however there are few published studies on equine response to essential oil odours. The study aimed to identify which essential oils were attractive to horses. The study comprised 10 horses (5 geldings, 5 mares) of mixed breed. Nine organic essential oils plus a control (no oil) were presented in a repeated measures experimental design. Oils were applied to cotton wool and gauze and hung either side of the stable door. Presentation of oils was randomised using a 10 x 10 Latin Square design. Oils were presented in pairs for 120 seconds with a 5 minute interval between tests. Horses were exposed to 5 paired presentations on day 1 and the experiment was repeated on day 2. Behaviour was recorded on videotape for 120 seconds. Continuous focal sampling was used to analyse behaviour based on an ethogram using Observer 5.0® software. The time the horses spent investigating each individual test substance was recorded. Range: 169.2 seconds (Violet Leaf) to 35.4 seconds (control). Kendall's W coefficient of concordance compared agreement in duration of interest between the horses. The mean ranked duration of investigation was significant ($W=0.272$, $n=10$, $p<0.05$). To compare differences between the duration of interest, individual oils were tested against the control using the Wilcoxon signed ranks test. Significant differences were found between the control and Peppermint (Mentha piperata) $p=<0.01$ and Violet Leaf (Viola odorata) $p<0.05$. Valerian (Valeriana officinalis) and Lavender (Lavendula angustifiolia) were approaching significance at $p=0.08$ and $p=0.09$ respectively. Although only a limited number of oils were tested, horses demonstrated a significant attraction to Peppermint, Violet Leaf and were also attracted to Valerian and Lavender. This study is relevant for horses exposed to essential oils in horse care products.

Recumbent behaviour in stabled horses depends on housing system

M. Van Straten, A. Lundberg and M. Andersson, Swedish University of Agricultural Sciences, Department of Animal Environment and Health, Box 234, SE-532 23 Skara, Sweden

The type of confinement influences the resting behaviour of horses and thus their welfare. Horses do sleep in slow wave sleep (SWS) while standing but have to lie down in a lateral position before entering Rapid Eye Movement sleep (REM). Studies in humans and other animals show that a lack of REM leads to psychological and physiologic problems. Since SWS cannot compensate for REM it is most likely that it is also necessary for horses to lie down in order to rest properly. The aim of this study was to investigate differences in recumbent behaviour between horses in tie-stalls and looseboxes. Twelve horses housed in looseboxes and tie-stalls in a stable in Sweden were video-recorded during night time in February and March 2007. Every horse was subjected to both treatments. The observations were paired within individual, as the horses were used as their own control. General Linear Model was used, called Mixed Procedure in SAS. Time budgets during the night (20h00 to 8h00) differed significantly between looseboxes and tie-stalls ($p<0.001$). In looseboxes horses were recumbent 2:15 hours per night compared to 1:29 hours in tie-stalls. When in lateral recumbency the average bout length was significantly longer in a loosebox compared to a tie-stall ($p=0.01$), which were on average 5:21 minutes and 4:17 minutes respectively. Studies of free-ranging horses show that they spent between 2.4% and 5.3% of their time budget recumbent compared to between 11% and 18% in this study. Possibly these 'wild' horses lay down only when necessary: when entering REM. However, horses in looseboxes spent the same amount of time in lateral recumbency as free-ranging horses, 2.6% of their total sleeping time (TST). Horses in tie-stalls though, spend only 1.4% of their TST in lateral recumbency and thus REM, and might suffer from sleep deprivation.

Evaluating social encounters of horses during mixing

E. Hartmann[1], J. Winther Christensen[2] and L.J. Keeling[1], [1]Swedish University of Agricultural Sciences, Animal Environment and Health, Box 7038, 750 07 Uppsala, Sweden, [2]University of Aarhus, Animal Health, Welfare and Nutrition, Box 50, 8830 Tjele, Denmark

Group housing of horses is not widely applied in practice despite clear welfare advantages. Concerns are raised regarding an increased injury risk or of how to introduce a new horse into an established group. This study investigated two hypotheses: 1) horses which meet in neighboring boxes before being put together in a paddock show a lower level of aggression compared to horses meeting for the first time 2) it is possible to predict the social encounter in the paddock after observing horses' behaviour in neighboring boxes. Two-year-old Danish Warmblood mares (n=20) were kept in two groups of 10 horses. In total, 60 encounters were arranged, whereby each horse was confronted pair-wise with six horses from the other group. There were two treatments: 1) box (B) + paddock (BP) and 2) only paddock (P). Each horse was confronted with three individuals according to each treatment. Horses met in the box for 5 minutes and in the paddock for 10 minutes. The frequencies of agonistic and non-agonistic behaviours were recorded continuously from video recordings. Results indicated no difference in aggression levels between BP (Mean ± SE; 8.13 ±1.06) and P (10.43 ±1.65; t_{57}=0.12, p>0.5). Noteworthy, the frequency of non-agonistic behaviours (e.g. friendly approach, nasal sniff) was higher in P (Mean ± SE; 10.53 ±1.18) compared to BP (6.2 ±0.77; t_{58}=3.10, p<0.003). No correlation was found between the aggression frequency in B and BP (r=0.3; ns). Allowing horses to meet in neighboring boxes had no impact on aggression when they were released into the paddock. However, as the horses exchanged non-agonistic 'greeting' type behaviours in the boxes these behaviours were reduced in the paddock. Further work is needed to clarify if increased pre-exposure times would affect aggressive behaviour during mixing.

Diurnal variation of plasma cortisol concentrations in stereotypic and non-stereotypic horses

K. Hemmann, M. Raekallio, K. Kanerva and O. Vainio, University of Helsinki, Faculty of Veterinary Medicine, Research Centre for Animal Welfare, Dept of Equine and Small Animal Medicine, Koetilantie 7, POBox 57, 00014 University of Helsinki, Finland

Crib-biting in horses is classified as an oral stereotypy, which has been suggested to functions to reduce stress. Irregular diurnal cortisol rhythm has been identified upon chronic stress in pigs and rats. Little information is available about naturally occurring patterns of 24-hour cortisol secretion in crib-biting horses. The purpose of this study was to determine the diurnal plasma cortisol concentrations in horses with stereotypic crib-biting. We hypothesised that the diurnal rhythm of crib-biters would be attenuated compared to their controls. Eleven warm blood horses or ponies, mean age 12.6 (6-20) years, 6 crib-biters and 5 controls, were included. All the horses lived in the same riding center and they were housed, fed and exercised following their everyday routines. Blood was collected via a catheter every 2 hours for 24 hours. Plasma was stored at -20 °C until analysed for cortisol concentration by radioimmunoassay. Student's t- test was used for comparisons between the groups for the diurnal maximum and minimum cortisol concentrations and the range of the actual plasma concentrations for each horse. The maximum cortisol concentrations (nmol/L, mean ± SD) were 143.4±22.7 and 148.5±58.1, (p=0.844) the minimum 43.7±15.8 and 45.9±25.3 (p=0.866) and their difference 99.6±18.6 and 102.6±44.9 (p=0.883) for crib-biters and controls, respectively. In conclusion, no significant differences were noticed in the diurnal plasma cortisol concentrations between crib-biters and non-crib-biters, but the individual variations were marked. A diurnal rhythm of cortisol was observed in all horses. The association between the diurnal cortisol variations and occurence of abnormal behaviour and stress should be further investigated.

Prevalence and risk factors for behavioural problems in leisure horses: how do they compare with those identified in performance horses?

J.S. Hockenhull and E. Creighton, University of Chester, Biological Sciences, Parkgate Road, Chester CH1 4BJ, United Kingdom

Owning horses primarily for recreation has grown over recent years and leisure horses now constitute the majority of the UK's equine population. However, the primary focus of much of the horse welfare research is on performance horses, made up of more reactive breeds and more intensively managed than their leisure horse counterparts. Consequently, the findings of these studies may lack relevance for leisure horses. A large-scale Internet survey of leisure horse owners in the UK was used to collect data on individual horses. Owners were asked to score the frequency their horse performed 11 different behaviour problems on a 1-5 scale. Seventy five percent of the 1223 horse sample reported some degree of behavioural problem. Of these, aggression towards horses (35%) and wood-chewing (34%) were the most common. Stereotypic behaviour was reported in 26% of horses, which is higher than any corresponding study on performance horses. This was possibly due to the use of rating scales which we have found promotes more accurate reporting than binary questions. Principle Components Analysis of the 11 behaviour items extracted four factors which together accounted for 63% of the variance in the data. These corresponded to inappropriate oral behaviour, aggressive behaviour, inappropriate locomotory behaviour and abnormal ingestive behaviour. The relationships between management practices and these factors were explored revealing risk factors such as the amount of time the horse spent stabled per day (locomotory χ^2_6=12.62, P=0.049; ingestive χ^2_6=29.83, P<0.001) and the type of bedding provided (oral χ^2_6=13.28, P=0.039; ingestive χ^2_6=148.06, P<0.001). There were similarities in some findings with data from performance horses, but others suggest risk factors unique to leisure horses e.g. experience of the owner. Overall, data indicate that prevalence of behaviour problems is high within the leisure horse population, and there are identifiable risk factors that could be tackled through owner education.

The effect of different handling treatments during repeated loading of unhandled British native ponies

H.A. Van De Weerd[1], S. Seaman[2], K. Wheeler[1], P. Goddard[3] and B. McLean[1], [1]ADAS UK Ltd., Gleadthorpe, Meden Vale, Mansfield, Nottingham, United Kingdom, [2]The University of Edinburgh, Easter Bush, Roslin, United Kingdom, [3]The Macaulay Institute, Craigiebuckler, Aberdeen, United Kingdom

Native ponies roam freely in natural habitats such as hills and moorland. They are rounded up annually to establish ownership and undergo welfare checks. This will be the first time they are in close contact with humans and may experience loading and transport to livestock markets. The objective was to study the effect of different handling treatments on the behaviour of 18 unhandled Welsh mountain ponies (18 fillies, <2-years old) during repeated loading. Ponies had no previous transport experience except when transported to the study site (one week earlier). During a 2-week period, animals were individually loaded 6 times into a stock trailer using a standardised procedure with two handlers. Three treatments were applied: low pressure (L), gentle voice, low arm movements (n=6); medium pressure; (M) gentle voice, gradual approach using handling boards (n=6); high pressure (H), loud voice, waving of arms and plastic bags (n=6). Ponies remained on the trailer for 10 minutes and their behaviour was recorded. Repeated measures analysis was used to establish treatment and time effects. Mean daily loading time was 16.2±1.3 seconds (mean ± SEM). L-ponies loaded significantly slower than other animals (L: 26.7±4.5, M: 13.1±0.7, H: 8.8±0.8 seconds, p<0.01). Unloading was quicker (overall daily mean: 4.7±1.3 seconds), but treatments did not differ. Loading time did not change over time, but unloading time decreased (time effect: p<0.01). Frequency of active behaviours (including movements) on the trailer did not differ between treatments (mean daily number 109.7±2.4), nor did it change over time. These findings suggest that the ponies did not habituate to repeated loading. The speed of unloading did increase, which could be a combined effect of learning the unloading procedure and motivation to rejoin the group. There were no differences between the M- and H-treatments, suggesting that loading with high pressure is not necessary for effective loading.

Morphological features as predictors of curiosity and reactivity in domestic horses

A.S. Brubaker and R.G. Coss, University of California, Davis, Psychology, 1 Shields Ave, Davis, CA 95616, USA

This study investigated the relationship of the position of heritable facial hair whorls – shown to correlate with reactivity in cattle – to reactivity, curiosity, and play behaviour in horses. Literature suggests that curiosity and reactivity to sudden stimuli are positively correlated, and a high whorl position indicates high reactivity. A pilot study tested 12 horses (6 mares, 6 geldings) of varying ages and breeds. Measurements included forehead height (nuchal crest to upper eye line), and whorl position (whorl center to upper eye line). Proportional Whorl Position (PWP) is the ratio of whorl position to forehead height. Positive values indicate the whorl is above the upper eye line; negative indicate the whorl is below it. An ANOVA run on a separate sample of 64 Thoroughbreds, Quarter Horses, and Arabians found no significant effect of breed on PWP: $F(2, 61) = 1.13$, $p=0.33$. Each horse was released into a 25-m arena containing a novel object (yoga ball) to assess spontaneous curiosity. Latency to make contact, and number and duration of bouts of interaction with the object were recorded for 10 minutes. A startle test consisted of a handler leading the horse past a rapidly opening umbrella, 1 meter away. Initial flight distance and latency to return within 1 meter of the umbrella were recorded. Analyses using STATISTICA ('98 edition) found that PWP was negatively correlated with duration of interaction ($r=-0.91$, $p<.001$) and number of bouts ($r=-0.60$, $p<.05$). Initial flight distance trended towards a positive correlation with PWP ($r=0.56$, $p=.056$), and a negative correlation with duration of interaction ($r=-0.52$, $p=.08$) and number of bouts (-0.56, $p=.06$). These preliminary results suggest that spontaneous curiosity may not be positively correlated with reactivity. Improved understanding of how observable morphological features may predict curiosity and reactivity behaviour has potential practical relevance to horse selection and training.

Behavioural assessment of nociception in awake Shetland ponies

M.C. Van Dierendonck[1], J.P.A.M. Van Loon[2], P.J. Stienen[3], A. Doornenbal[3], M.J. Van Der Kraats[1] and L.J. Hellebrekers[2], [1]Faculty of Veterinary Medicine; Utrecht University, Animals, Science and Society, Yalelaan 2, 3584 CM Utrecht, Netherlands, [2]Faculty of Veterinary Medicine; Utrecht University, Equine Science, Yalelaan 114, 3584CP Utrecht, Netherlands, [3]Faculty of Veterinary Medicine; Utrecht University, Clinical Sciences of Companion Animals, Yalelaan 108, 3584 CM, Utrecht, Netherlands

There are few tools for the objective quantification of nociception in the horse, neither physiologically nor behaviourally. This study was part of a larger project aiming to assess reliable (neurophysiological) measurements of (anti)nociception in the horse, since horses do not show nociception reliably. The ultimate project aim is better assessment of nociception to optimise analgesia and thus the welfare of horses. Seven awake Shetland ponies were subjected to electrical stimulation by means of electrodes placed on a distal hind limb. The stimuli were square-wave electrical stimuli of 5 ms duration, with an intensity ranging between 0.2-4 mA, 32 times per intensity. Following each stimulation, the nociceptive/aversive response was recorded by standardised behavioural observations from video. Based on the results from stimulus intensity-response sessions, relevant behavioural parameters were determined. Ten parameters were identified and subsequently clustered into two distinct response patterns: behavioural ipsilateral-reflex-score (IRS) and behavioural body-score (BoS). The IRS and BoS increased significantly with increasing stimulus intensity ($F_{(5,20)}=41.16$, $p<0.001$). There was no significant difference between the two scores ($F_{(1,4)}=0.35$, $p=0.58$), which allowed clustering of both scores in the next phase of the experiment (Total-Behavioural-Score: TBS). Subsequently, with two weeks interval, either an opiate (methadone, 0.5 mg/kg) or saline, was administered epidurally and recordings were performed every 5 minutes for a 40 minutes period. Next an opiate antagonist (naloxone, 0.04-0.06 mg/kg) or saline, was administered and recordings were continued every 5 min for another 40 minutes. Analysis showed that the TBS was affected differently by saline/saline and methadone/naloxone treatment over time (treatment*time: $F_{(3,18)}=9.35$, $p=0.001$). Post-hoc analysis showed that, in contrast to in the saline group ($F_{(3,18)}=3.12$, $p=0.52$), the TBS significantly decreased after methadone injection (time: $F_{(3,18)}=3.58$, $p=0.035$). Significant differences between saline/saline and methadone/naloxone treatments were found in three out of four compared periods. This behaviour classification will allow for more objective animal based assessments of nociception-related responses to be carried out in future experimental stimulation studies.

Dairy farmers assessment of animal pain, using a photo-based instrument, can predict animal welfare outcomes at farm level

C. Kielland, E. Skjerve, O. Østerås and A.J. Zanella, Norwegian School of Veterinary Science, P.O. box 8146 Dep, 0033 Oslo, Norway

Differences in welfare outcomes at farm level are influenced by the quality of human-animal interactions. We hypothesised that the attitude of farmers towards animal pain is reliable predictor of the quality of human-animal interactions and as a consequence is a good predictor of welfare outcomes at farm level. Photos are powerful in eliciting measurable empathic responses to pain in humans. We developed a photo-based pain assessment instrument (PAI) to assess pain in dairy cattle. Our goal was to test whether attitudes towards pain, measured using the PAI, influenced animal welfare outcomes at the farm level. We also measured the impact of demographic factors on the pain assessment score in individual farmers. Farmers (n=221) who participated in an epidemiological study received the PAI via e-mail. They were asked to asses how painful 25 different conditions were, marking on a 10 cm visual analogue scale (VAS). The response rate was 70% (n=154). The three conditions ranked as the most painful were a) fracture of tuber coxae, b) dystocia and c) serious mastitis. Regression analysis on the median score on all conditions for each farmer showed that for each 10 years of experience in dairy farming, farmers scored 0.24 lower in their median pain assessment (p=0.042). Interestingly for each of the 25 conditions they had actually experienced at the farm, they scored 0.13 higher (p=0.002). Additionally for each 1000kg increase of milk production/cow/farm, they scored 0.2 lower (p=0.059). According to the overall response pattern, cluster analysis (complete linkage) revealed 3 distinct groups. There were significant differences between the 3 groups with regards to age, milk production and a farm reproductive index. These analyses indicate that these factors influenced how pain is recognised and assessed by farmers, and might therefore influence the overall welfare at farm level.

Rats in school!

N.H.F. Franco, J.B. Santos and I.A.S. Olsson, IBMC - Instituto de Biologia Molecular e Celular, Rua Campo Alegre 823, 4150-180 Porto, Portugal

For an institution developing research mainly on the cellular and molecular level, finding attractive outreach activities for a very young public is challenging. We made use of an otherwise controversial topic - laboratory rats - to create a primary school project that took the well known concept of the classroom pet and gave it a little twist to increase its scientific value. In this project, fourth-grade children use the scientific method to study the behaviour of two female Lister-Hooded rats, as well as design all (non-invasive) experiments to discover more about the way the animals react to new experiences. Three classes in two different schools house one pair of rats per classroom, in large and specially designed habitats, equipped with dark-vision cameras filming the rats 24/7. Each habitat contains one part where the animals live and a separate part where the children can set the rats new challenges: mazes, different objects, food or smells, for example. The children use their own questions about the animals' behaviour as starting point for experiments designed in interaction with the teacher and a science communicator. During the two years of this pilot project, we have observed how the dynamics between science, scientists and schools have contributed to the development of the children's capacity for reasoning and critical thinking as well as their interest in understanding animal behaviour. A formal evaluation is underway, and the preliminary observations indicate that as the children have gained experience, their questions have become more informed and complex: rather than wanting to "see the animals doing something", they ask "what the animals do" in a particular situation and "why".

How do rats perceive music and noise

T.C. Krohn[1], B. Salling[2], S. Velschow[3] and A.K. Hansen[1], [1]University of Copenhagen, Faculty of Life Sciences, Department of Veterinary Pathobiology, Centre for Applied Laboratory Animal Research, Dyrlægevej 88, 1st floor, DK-1870 Frederiksberg C, Denmark, [2]Scanbur A/S, Silovej 16-18, DK-2690 Karlslunde, Denmark, [3]H. Lundbeck A/S, Ottiliavej 9, DK-2500 Valby, Denmark

An unstudied issue that is widely discussed is the effect of a radio playing in the animal facility. In some facilities, this is common practice, but depending on the volume of the radio, and possibly the program (music, talk-shows, or classical music) it could be expected that this may affect the animals in some way. In the present study, the rats' ability to distinguish between different kinds of music or noise was tested at a volume of 60 dB(A). In a preference test, ten rats (HsdOla:LH) had to choose between different cages with different music or noise placed on a digital weight with continuous sampling. Depending on the weight of the cages, and thereby on the choice of cage, it could be determined, whether the rat was able to distinguish between the music in the different cages. Data were tested for normal distribution and significance between the different choices was tested with t-test. The results indicate that the animals are able to distinguish between noise, music and radio. Uniform noise, such as white or pink noise, does not seem to disturb the rats, while structured sound or noise is not selected by the rats. Dwelling time per cage during the day (rest-period) in percent: Silence vs White Noise (63.7±33.2 vs 36.3±33.2), Silence vs Mozart K448 (57.2±38.6 vs 42.8±38.6), White Noise vs Mozart K448 (73.1±35.1 vs 26.9±35.1)*, Silence vs Radio (81.8±22.1 vs 18.2±22.1)***, White Noise vs Radio (80.5±20.6 vs 19.5±20.6)***, White Noise vs White Noise with speak (51.8±38.0 vs 48.2±38.0). It seems that the composition of the sound-pattern is important and the rats are able to distinguish between the different sound-patterns.

Does breeding of laboratory rats in larger cages affect behaviour, corticosterone and blood pressure?

K. Cvek[1] and L. Arborelius[2], [1]Swedish University of Agricultural Sciences, Department of Clinical Sciences, P.O. Box 7054, 750 05 Uppsala, Sweden, [2]Karolinska Institutet, Department of Clinical Neuroscience, Section for Psychology, 171 76 Stockholm, Sweden, Sweden

This study aimed at investigating effects of housing laboratory rats in larger cages on behaviour and physiology during the lactation period, in both dams and pups. Telemetric transmitters for registration of blood pressure, heart rate and activity were surgically placed in female Wistar rats, subcutaneously on the flank with the femoral artery catheterised. When fully recovered, the rats were mated and randomly placed singly in standard cages (ST; 2240 cm^2, height 20 cm; n=8) or standard cage bottoms with raised tops (RT; height 50 cm; n=7) fitted with a shelf. Black plastic tubes and nesting material was placed in all cages. Maternal behaviour was studied through direct observations once per minute for 90 minutes at five different occasions during the lactation period. Blood pressure, heart rate and activity were measured telemetrically at postnatal day (PND) 12. At weaning, urine samples were collected by placing the animals in empty cages. The urine was analysed for creatinine/corticosterone ratio. Preliminary results indicate no differences in the amount of time the dams spent with their pups, but the ST dams displayed more arched-back nursing and licking/grooming of the pups than dams in RT cages (data from all animals not yet analysed). This may imply that RT dams had not adjusted fully to the new cages. No differences in blood pressure, heart rate or activity of the dams could be detected during PND 12 between cage types or between day and night. Nor could any differences in creatinine/corticosterone ratio be detected between the dams (RT=75±47SEM, ST=145±89, t-test p=0.157). All pups were very active during the second half of the lactation period. However, the pups in the RT cages had more space and opportunities to climb and perform physical activity and also had significantly lower creatinine/corticosterone ratio than the pups in the ST cages (RT=416±61SEM, ST=638±53SEM, Mann-Whitney Rank Sum Test, p<0.001) indicating a lower stress level.

Auditory enrichment for zoo-housed Asian elephants (*Elephas maximus*)
D.L. Wells and R.M. Irwin, Queens University Belfast, School of Psychology, School of Psychology, Queens University Belfast, Belfast, BT7 1NN, United Kingdom

Auditory stimulation has long been employed as a method of therapy for humans and animals housed in institutions. This study explored the effect of auditory stimulation in the form of classical music on the behaviour and welfare of zoo-housed elephants, a species that is notoriously difficult to maintain successfully in captivity, and is much in need of welfare attention. Four zoo-housed female Asian elephants, *Elephas maximus*, were exposed, in an ABA design, to 2 conditions of auditory stimulation: a control (no auditory stimulation), and an experimental condition, during which the animals were presented with a commercially available CD of classical music. Each condition lasted for 5 days, with an interim period of 2 days between each condition. The elephants' behaviour was recorded using instantaneous scan-sampling every minute for 4 hours a day for the full five days of each condition, resulting in 240 scans per elephant per day. A Friedmann ANOVA was conducted for each behaviour to determine whether it was influenced by the auditory environment. Analysis revealed that the elephants, as a group, spent significantly (χ^2=8.00, df=2, P=0.01) less of their time stereotyping during the experimental condition (mean percentage of daily scans=6.97 +/-6.54) than both the pre- (mean percentage of daily scans=43.43 +/-19.51) and post- (mean percentage of daily scans=54.06 +/-19.25) control conditions. None of the other behaviours recorded were significantly influenced by auditory stimulation (P>0.05 for all Friedmann ANOVAs). Overall, the findings suggest that auditory stimulation in the form of classical music may be a useful method of reducing stereotypic behaviour in zoo-housed Asian elephants, although more long-term work with a larger number of animals is needed before firm conclusions can be drawn.

Do detailed measurements of sleep and rest tell us more about animal welfare than simple measures of lying behaviour?

F.M. Langford[1], K.M.D. Rutherford[1], M.J. Haskell[1] and M.S. Cockram[2], [1]SAC, Sustainable Livestock Systems, West mains Road, Edinburgh, EH9 3JG, United Kingdom, [2]Atlantic Veterinary College, Sir James Dunn Animal Welfare Centre, University of Prince Edward Island, 550 University Ave, C1A 4P3, Charlottetown, Canada

In many animal behaviour and welfare studies, the time an animal spends resting and sleeping is counted together to form a behavioural category of 'inactivity'. An underlying implication is that 'inactivity' is not particularly important to study as opposed to 'activity' –including everything else from feeding, playing and socially interacting. However, animals are motivated to rest and to sleep and most animals spend a large percentage of their time 'inactive'. We know that the time spent 'inactive' can be affected by the environment, but often lack of detail in recording 'inactivity' results in the identification of only a few causational factors. For example, in a study investigating the welfare of dairy cows at farm level, there were positive correlations between lying behaviour and both cubicle comfort and lameness. From measuring the amount of lying alone, it was not possible to ascertain whether cows were lying because they were comfortable, or because they were in pain when standing. Undertaking detailed behavioural measurements of resting and especially sleep may help us to understand how animals are affected by their environment. In humans, resting behaviour and sleep quantity, quality and patterns are altered by pain, illness, stress, learning, positive feelings, depression and fatigue. Many of these relationships have been demonstrated in laboratory animals as models for human sleep disorders. Recent research presented at ISAE on the effect of husbandry procedures on rest and sleep in sheep (Langford *et al.*, 2004) and calves (Hänninen *et al.*, 2006) indicates that these relationships may also exist in farm animals. Detailed measurements of rest and sleep are potentially useful tools to assess farm animal welfare.

Resting posture and posture changing behaviour as indicators of floor comfort for finishing pigs

K. Scott, A.J. Balsom, J.H. Guy and S.A. Edwards, Newcastle University, School of Agriculture, Food & Rural Development, Agriculture Building, Newcastle upon Tyne, NE1 7RU, United Kingdom

Pigs spend ~80% of their time resting, therefore floor comfort and suitability are important welfare concerns. This study investigated suitability of resting posture and posture changing behaviour as indicators of floor comfort. Four replicate pens of 8 pigs were studied between 70 and 100 kg on each of three floor types: part-slatted concrete (PS), fully-slatted concrete (FS) and deep straw bedding (ST). Pigs were assessed (0-5 scales) for leg lesions and adventitious bursitis. Lying behaviour was recorded over 24-hour periods, using time-lapse video recordings, to give 10 minute sampling of the number of pigs adopting a sternal, lateral, semi-lateral, standing and sitting position. A video camera was used to take real time records of pigs during posture changing between standing and lying, and lying and standing. The effects of floor type on posture frequency, lesion scores and transition timings were analysed by ANOVA. Pearson's correlations were used to assess relationships between lesions and lying postures. ST pigs had lower hind leg lesion scores (ST=0.6, FS=1.8, PS=1.2, S.E.M.=0.06, $F_{2,9}$=91.04; P<0.001), whilst hind limb bursitis was more severe in FS pigs (FS=0.7, PS=0.02, ST=0.02, S.E.M.=0.114, $F_{2,9}$=13.19; P<0.01). At 100 kg, FS pigs were more frequently observed in a sternal lying posture (FS=22.4, PS=10.3, ST=11.9%, S.E.M.=2.71, $F_{2,9}$=5.90; P<0.05). Lying posture frequencies did not correlate to lesion and bursitis scores. Pigs on FS floors took longest to complete the standing to lying transition (FS=5.43, PS=4.34, ST=3.06 sec, S.E.M. =0.276, $F_{2,9}$=18.46; P<0.001). Lower leg (r=0.863, P<0.01) and hock bursitis scores (r=0.589, P<0.05) were significantly correlated with the total duration of this transition. From relationship to injury, posture changing behaviour therefore seems a more useful measure of the adequacy of floor types for pigs than time in different resting postures.

Does a missed milking affect cow lying behaviour and comfort?

K. O'Driscoll[1], P. Gazzola[2], L. Boyle[3], D. Gleeson[2] and B. O'Brien[2], [1]Centre for Animal Welfare and Anthrozoology, Department of Veterinary Medicine, Cambridge, CB3 0ES, United Kingdom, [2]Teagasc, Moorepark Dairy Production Research Centre, Fermoy, Co. Cork, Ireland, [3]Teagasc, Pig Development Unit, Fermoy, Co. Cork, Ireland

Omission of one evening milking per week has lifestyle and labour advantages for dairy farmers. This study aimed to evaluate the effect of 13 times weekly milking on cow comfort, using dairy cow lying behaviour as an indicator. Spring calving cows (n=24) (mean calving date: 20 February 07±17days) were randomly assigned to two treatments - milked either 14 (14W) or 13 (13W) times weekly. 13W cows were not milked one evening each week. Standing/lying was recorded every 5 min using modified voltage dataloggers (Tinytag Plus, Chichester, UK) over 3 days on 2 consecutive weeks (beginning 73±19 DIM). Recording commenced the day prior to (d1), and finished the day after (d3) the day that evening milking was omitted (d2). Udder tension (scored 1-3) and milk leakage (yes/no) were recorded the morning of d2 and d3 for three consecutive weeks (beginning 73±19 DIM). Lying behaviour was analysed using mixed models. Udder tension and milk leakage were analysed using logistic regression and chi-square tests. Lying behaviour was similar in both treatments on d1 and d3. There was an interaction between treatment and day for total lying time (F=9.65, P<0.001) and lying bout duration (F=3.36, P=0.01). On d2 13W tended to have shorter lying bouts than 14W cows (approx 47 vs. 70 min; T=2.49, P=0.1). On d2, 13W cows tended to spend less time lying/hour than 14W cows, particularly 13 to 24 hours post milking (T=2.78, P=0.06). On d3, 13W cows had higher udder tension scores (P<0.001), and greater incidence of milk leakage (P<0.001) prior to milking than 14W cows. Although changes in behaviour were transient, omission of evening milking resulted in shorter lying bouts and reduced the overall time spent lying. This was probably due to udder discomfort, considering udder tension and milk leakage results. Thus lying behaviour may provide insight into udder-related discomfort in cows.

The use of an activity monitor to assess the welfare of dairy cattle

N. Blackie, E.C.L. Bleach, J.R. Amory and J.R. Scaife, Writtle College, Centre for Equine and Animal Science, Lordship Road, CM1 3RR, United Kingdom

Lying behaviour has the potential to evaluate the welfare of cattle by identifying lame cows. Lying surface quality could be predicted by the frequency of posture changes and presence of hock lesions. The following studies are towards the creation of an automated lameness detection system. These studies were conducted using IceTag™ activity monitor on a 500 cow dairy farm. Data obtained includes time spend lying, standing and active; also length, distribution and frequency of lying bouts. Analysis of variance was used for studies 1-3 and regression analysis was used in study 3, using Genstat. Study 1: Lame (n=22) cows spent significantly longer lying down (p<0.05) than non lame (n=37) cows (mean ± SED 13.0 vs. 11.0±0.74 hours). Lame cows also had longer average and maximum lying bout length than non lame cows. Study 2: Cows in parity 1 (n=10) spent significantly less time (p<0.05) lying down than cows in parity 3+ (n=28) (9.9 vs. 12.2±2.35 hours). Stage of lactation can also impact on lying behaviour with cows in week 6 (n=7) of lactation spending significantly less time (p<0.05) lying down than week 32 of lactation (9.0 vs. 12.3±1.79 hours, respectively). This difference may be due reduced time spent feeding in late lactation. These cows also showed a significantly longer average lying bout in week 32 compared to week 6 of lactation. Study 3: Frequency of lying bouts throughout the day can give an indication of cubicle comfort, with cows with hock lesions greater than 4cm (n=10) lying down more frequently (p<0.05) than cows with lesions 0-3cm (n=31). There was a significant positive relationship between hours spent lying down and the number of hock lesions seen (r^2=11.7, p<0.05). Number of hock lesions was not related to locomotion score; however cows with locomotion score 3 had larger hock lesions (p<0.05) than locomotion score 1 cows (2.6 vs. 3.6±1.15cm)

An holistic analysis about the effectiveness of environmental enrichment programmes: study cases in the Chilean national zoo

G. Cubillos, Zoologico Nacional del Parque Metropolitano de Santiago, Zoo, Pio Nono, 450, 8420000, Chile

In 1998 the Chilean National Zoo started to introduce Environmental Enrichment (EE) to captive animals assuming it would have a positive effect on their welfare. However, these procedures lacked scientific evidence that would demonstrate such effects and validate their application. Since 2002, we began to apply a more systematic approach: different programmes were developed for specific groups of animals aimed at their particular behavioural needs and behavioural records were collected. This paper includes the cases of the carnivore order represented by the three families and seven species present in our collection: felids, (Panthera onca, 1.1.0; Panthera tigris, 1.1.0; Puma concolor, 1.2.0, and Panthera leo, 1.5.0), ursids (Tremarctos ornatus, 1.0.0, and Ursus arctos, 1.2.0) and canids (Pseudalopex culpaeus, 1.1.0). The approach and analysis in each case depended on the nature of data and the aims of each study. However, for the purpose of this analysis each case is evaluated according to the effectiveness of each programme and differential effects within a group of individuals. In general, the results for the three families showed a significant increase in foraging, manipulation, interaction with EE items, use of space, and a decrease in pacing stereotypes, when present (Mann-Whitney U test, $p < 0.05$). Systematic observation also evidenced behavioural habituation and dishabituation to items. The success of EE programmes will be discussed suggesting it depends on the frequency and variability of application. Potential differential effects within a group depend on individual history and can be addressed with specific routines. New methods to objectively quantify behavioural response to EE stimuli, estimating potential benefits for the animals will be presented.

Assessing effects of regular environmental enrichment immediately and after a month of presentation: implications for welfare of Cape Clawless Otters (*Aonyx capensis*)

M.K. Sharra[1], B. Stark-Posta[2] and J.M. Siegford[1], [1]Michigan State University, Animal Behaviour and Welfare Group, Department of Animal Science, 1290 Anthony Hall, East Lansing, MI 48824, USA, [2]The Toledo Zoo, P.O. Box 140130, Toledo, OH 43614, USA

Environmental enrichment is frequently used with a wide range of exotic species in captivity. Provision of enrichment is assumed to be beneficial, but little is known about the long-term effects on behaviour and therefore on welfare. This study was conducted to provide insight into behavioural responses to enrichment with respect to immediate and longer term changes in activity. The behaviour of 2 cape clawless otters (*A. capensis*) at the Toledo Zoo was compared prior to the regular presentation of enrichment, immediately, and one month after adding regular enrichment as part of a 2007 summer internship study undertaken by M. Sharra. Data from past behavioural studies on similar species were used to predict how enrichment should affect the otters in this study. Enrichment items were chosen to stimulate natural behaviours of the species such as digging, foraging and swimming and to encourage a wider range of exhibit utilisation. Adding regular enrichment resulted in immediate increases of locomotion and enrichment-oriented behaviours however, longer term rates of activity were similar to baseline levels. Similarly short-term results of changes in use of locations in the exhibit were observed which were related to the locations where increased enrichment-oriented behaviours occurred. Immediately after incorporating enrichment, social interaction between the otters increased but, again, longer interaction rates dropped back to baseline rates. While the addition of regular enrichment does increase desired behaviours and exhibit interaction, the effects in this study were short-lived, indicating the importance of long-term monitoring. To maintain the effects of novelty, increasing the diversity of enrichment items and presentation may effect more long-lasting behavioural changes. Providing a wide variety of species-appropriate items will likely increase enrichment-oriented behaviours however, a systematic study of the long-term effects of enrichment is needed for a wide variety of species in order to develop practical yet effective enrichment strategies.

Effects of human interaction on the behaviour of a colony of common marmosets (*Callithrix jacchus*)

A. Manciocco, F. Chiarotti and A. Vitale, Istituto Superiore di Sanità, Department of Cell Biology and Neuroscience, Viale regina Elena, 299, 00161, Italy

In order to maximise the well-being of captive animals every aspect of their life should be attended to. Human interaction as environmental enrichment for non-human primates is believed to be of value, but so far it has been subject to little quantitative evaluation. This study assessed the effects of positive human interaction on the behaviour of a common marmosets' (*Callithrix jacchus*) colony housed in laboratory conditions. Experimental subjects were 12 individuals housed in three groups (2, 4, 6 individuals respectively). The study comprised three consecutive phases. In the baseline phase, the caretaker carried out routine duties in the monkeys' room and subsequently data collection began; in the second phase, the caretaker spent additional 20 minutes per day in the room, interacting with individuals (no data collection); in the third phase, identical procedure for human interaction was carried out and data were collected. This began 30 minutes after the caretaker had left. The experimental unit was the individual and the sampling method was a 30s focal animal scan sampling, with daily 30-min. long sessions. An ANOVA was used and the Tukey post hoc-test was performed. Following the period of unstructured interaction, the marmosets showed increased grooming levels ($F_{1,7}=11, 46, P=0.002$) and playful activities ($F_{1,7}=8, 49, P=0, 0004$), while lower levels of self-scratching ($F_{1,7}=28.4, P=0.001$) and locomotion ($F_{1,7}=5.41, P=0.05$) were found. Furthermore, a decrease of contact vocalisations ($F_{1,5}=10, 33, P=0,004$) and a trend towards a reduction of agonistic displays to the observer ($F_{1,7}=4, 17, P=0, 08$) were recorded. These results suggest that unstructured positive interactions between familiar humans and marmosets can influence the behaviour of the animals and that this should be taken into consideration in the management of captive non-human primates.

To capture the interest of student: the means of introducing applied ethology and animal welfare in Romania

B. Algers, Swedish University of Agricultural Sciences, Department of Animal Environment and Health, Gråbrödragatan 19, SE-53223 Skara, Sweden

In 1994 I employed a Romanian veterinarian, Mrs Maria Onila, as a researcher at my department. As she became aware of the activities and competence that we had at our department, she wanted to make connections with her home country to expose the staff and students of the veterinary faculty in Timisoara. I decided to use clinically interesting examples to show how behaviour links to things that were already well known for these staff and students. By using an interactive approach to the auditorium, who were used to lectures with little or no interaction with the audience, I could attract the students' interest. As a result I was invited several times to come back to give new lectures to veterinary medicine students in Timisoara and Bucharest, and a small project collaboration was developed. I also helped to provide scientific material which was difficult for the faculties to get hold of. The result was that my colleague in Romania, Professor Michai Decun, wrote a book, Etologia, bunastarea si protectia animalelor, the first text book in Romanian on ethology, welfare and animal protection. Prior to 2001 animal welfare did not exist as a taught discipline in any of the faculties in Romania with a biological or agricultural profile. In 2001 a course Ethology, welfare and animal production with 42 course and 28 training hours was introduced in Timisoara. In 2003 and 2004 the faculties in Bucharest, Cluj and Iasi introduced courses in Animal Welfare and Protection. Future steps must include the exchange of PhD students between faculties in Romania and those universities who have lots of experience in postgraduate training in applied ethology and animal welfare. And due actions must be taken to ensure that the teaching and research activities in applied ethology is founded on a thorough knowledge of basic ethology.

Important aspects for the development of applied ethology in Brazil and elsewhere

L.C. Pinheiro Machado, Universidade Federal de Santa Catarina, Department of Zootechny, Rod. Admar Gonzaga, 1346 - CCA/UFSC, 88.034-001 - Florianopolis, SC, Brazil

A key factor for the development of Applied Ethology (AE) in Brazil was the existence of people, in different regions of the country, interested in and committed to its development. For this group to gain ground, a relatively open scientific community, generating research and teaching opportunities, was necessary. Two major events were, in 1981, the creation of the first course in the Agronomy curriculum at the Federal University of Santa Catarina (UFSC) and, two years later, the First National Meeting on Ethology in the São Paulo State University (UNESP). Since then, annual nationwide meetings have been held, and in 1992, the Brazilian Society of Ethology was finally created. As a result of the growing involvement of scientists with ISAE, in the year 2000, the 34th International Congress was organised in Brazil. The ISAE was very important, supporting researchers and the launching of a Congress, it gave further impulse to the field. AE is, nowadays, present in the curricula of most undergraduate courses in Agriculture and is part of many Graduate courses. AE in countries like Brazil and Mexico was further boosted by long lasting two-way collaborations with ISAE scientist members from around the World. However, Brazilian ethologists were initially denounced among the local scientific and academic community for their "romantic" views, as some would argue that "animals don't really have feelings", AE "is animal management" or "is already covered by Animal Physiology". The supportive attitude of ISAE was greatly important to overcome these obstacles. In our opinion, the best way for ISAE to promote AE worldwide is giving opportunities to scientists of countries with less tradition: grants to attend Congresses; organisational opportunities of Congresses (ex: Brazil and Mexico); publishing opportunities (e.g. AABS). Being open to reflect a diversity of realities makes ISAE truly international, attractive and exciting society for new members.

Taking up the baton in applied ethology and animal welfare education worldwide

M.J. De Boo, World Society for the Protection of Animals (WSPA), Programmes Department, Education Unit, 89 Albert Embankment, London SE1 7TP, United Kingdom

While the results of applied ethology research may be applicable to animals across borders and societies, introducing the principles where only a basic concept of animal welfare exists, can be complicated. Using applied behaviour science effectively requires knowledge of local customs, culture, politics, and curriculum development, and a tailored approach to a specific audience. The influence of the veterinary profession on animal welfare is important due to their potential to influence, and implement, strategies at local, national and international level. The World Society for the Protection of Animals (WSPA) Concepts in Animal Welfare (CAW) syllabus is a tool used to aid the teaching of animal welfare in veterinary curricula worldwide. International (e.g. OIE, CVA, WVA) and national (Brazil, Philippines) veterinary associations and bodies have endorsed the syllabus as a useful resource. Experiences gained from four years of introducing this resource to faculties globally, demonstrates some of the needs, challenges and solutions found while working towards a single objective in a variety of cultures and countries. Successful inclusion of animal behaviour and welfare into veterinary curricula varies depending on the country or region, and the individual motivation of University officials. Case studies of implementing CAW in Brazil, Kenya, Mexico and the Philippines highlight these varying motivations, challenges and successes. Successful approaches such as CAW in veterinary education could be adopted in similar academic disciplines such as agriculture/animal science, and even social sciences. Perhaps the ISAE could take the lead in initiating a 'generic' applied ethology syllabus.

Selection and welfare: using breeding tools to improve welfare

M.J. Haskell, S.P. Turner, R.B. D'Eath, R. Roehe, E. Wall, C.M. Dwyer, E.M. Baxter, J.M. Gibbons, G. Simm and A.B. Lawrence, SAC, Sustainable Livestock Systems Group, West Mains Road, Edinburgh EH9 3JG, United Kingdom

Breeding programmes that focus exclusively on production often have unforeseen side-effects on traits related to health or fitness. In response, more of these traits are being included in selection programmes. Additionally, it would be beneficial to animal welfare to include behavioural traits, such as good maternal behaviour, calm responses to handling and reduced aggressiveness. Currently, there are few working examples of this. To include behaviour in selection, it must be possible to reliably measure the trait of interest in individual animals. As many animals must be assessed, the trait must be measurable quickly and/or automatically. To overcome this, proxy measures may be developed and validated e.g. lesion scoring as a proxy for mixing aggression in pigs. In the future, molecular genetic markers may be used. Then, sufficient animals with a known pedigree must be measured to calculate heritability (proportion of observed variation due to additive genetic effects). This is used to derive individual estimated breeding values (EBV: estimate of the animal's genetic merit). Currently, EBVs are available for docility in beef cattle and 'temperament' in dairy cows. EBVs for traits such as pig aggressiveness at mixing and lamb vigour are under development. EBVs are valuable tools that breeders can use in selecting elite breeding stock or culling animals with very negative values. Breeders often use selection indices to weight emphasis on several EBVs (e.g. production and health) to maximise response in the overall breeding objective (e.g. profitability). It may be desirable to include behavioural traits, directly or indirectly, in future. This requires estimates of genetic correlations among the traits. A profitability index also requires economic values for each trait. However, many welfare traits have value beyond their associated costs or market value. Alternative methods are needed to integrate these traits into indices, and to reward breeders for using these traits.

Searching for genes involved in feather pecking behaviour in laying hens

A.J. Buitenhuis, University of Aarhus, Department of Genetics and Biotechnology, Blichers alee 20, DK-8830 Tjele, Denmark

Feather pecking (FP) is a major problem in the laying hen industry. It is characterised by pecking at- or pulling out of feathers of the other animals and results in denuded areas. Eventually FP may result in cannibalism. Selection lines selected for 10 generations for reduced (LP) and increased (HP) pecking behaviour, differing significantly in their pecking behaviour, are available. A C(α) test for mixtures in the distribution of number of pecks has shown that there is a major dominant allele associated with very high FP in the HP population, which makes the selection lines an ideal resource to study the genetic mechanisms underlying FP. In this experiment 60 animals of the 8th generation from both the LP line and HP line were kept in groups of 20. The pens were 2 m x 4 m with wood-shavings on the floor. Feed and water were provided ad libitum. All pecks to other chickens were recorded by video from 14:00 to 17:00. The experiment has been performed according the regulation of the Danish Committee of Control with Animal Research (Dyreforsøgstilsynet). The next day the birds were decapitated. The brain was frozen in liquid nitrogen and stored at -80C. The 20K chicken oligo array (ARK genomics) was used for gene-expression profiling on whole brain tissue of the LP and HP hens. The contrast between the LP and HP hens revealed 40 differentially expressed genes at the FDR adjusted_P_value <0.10 level (LIMMA version 2.10.0 under R version 2.5.0). As the annotation of the genes still needs to be improved, interpretation of the function of these genes is difficult. Nevertheless, if confirmed in further studies these genes and their expression patterns may contribute to a better understanding of pecking behaviour in laying hens.

Maternal infanticide in pigs

C.R. Quilter[1], S.C. Blott[2], A.E. Wilson[1], M.R. Bagga[1], C.A. Sargent[1], G.L. Oliver[1], O.I. Southwood[3], C.L. Gilbert[4], A. Mileham[1,5] and N.A. Affara[1], [1]University of Cambridge, Pathology, Tennis Court Road, Cambridge, CB2 1QP, United Kingdom, [2]Animal Health Trust, Lanwades Park, Kentford, Newmarket, CB8 7UU, United Kingdom, [3]PIC, 2 Kingston Business Park, Kingston Bagpuize, OX13 5FE, United Kingdom, [4]The Babraham Institute, Babraham Hall, Cambridge, CB22 3AT, United Kingdom, [5]Genus plc, 1525 River Road, DeForest, WI 53532, USA

The aim of our study was to identify genes and/or chromosomal regions associated with maternal (infanticidal) sow aggression, defined by sows attacking and killing their own newborn offspring, within 24 hours of birth. This is likely to be a multifactorial phenotype where there is interaction between multiple genes predisposing to an aggressive phenotype ameliorated by experience and the environment. We also propose that this could be an animal model for puerperal psychosis, the severe form of human post-natal depression. A dual approach was carried out: A whole genome linkage analysis using 80 microsatellite markers on 121 sib pairs of pigs and a microarray approach where hypothalamus samples from 9 matched pairs of aggressive/control animals were hybridised to a pig array made from hypothalamus cDNA libraries. The affected sib pair analysis identified 4 quantitative trait loci (QTL) mapping on Sus scrofa chromosomes 2 (SSC2), 10 (SSC10) and two on X (SSCX). Microarray analysis identified 74 priority genes as being differentially expressed in the infanticide phenotype. Identification and understanding of the maternal infanticide phenotype is of utmost importance to the agricultural community with the ultimate aim of putting in place measures to reduce maternal aggression in the sow. This will be important in terms or animal welfare by reducing the number of savaged piglets and in addition reducing the stress exhibited by the mother herself when exhibiting this aberrant behaviour. We are currently looking for single nucleotide polymorphisms (SNPs) across the range of our QTL regions and also in our array candidate genes for the purpose of refining our QTL and developing a predictive test. Any confirmed associations to the maternal infanticide phenotype could be extended to an analysis of the orthologous genes in patients with puerperal psychosis.

Domestication effects on social behaviour in chickens is partly determined by genotype on a growth QTL

A. Wirén[1], U. Gunnarsson[2], L. Andersson[2] and P. Jensen[1], [1]Linköping University, IFM Biology, SE-581 83, Linköping, Sweden, [2]Uppsala University, Department of Medical Biochemistry and Microbiology, Box 582, SE-751 23 Uppsala, Sweden

During domestication, animals adapt to a life with humans, often in large social groups. We studied social behaviour towards unfamiliar conspecifics of wild and domestic chickens, and also compared chickens differing in genotype on one locus only. This could shed light on the genetic mechanisms of behavioural adaptation during domestication. We performed two experiments in chickens from an advanced intercross line (AIL) between White leghorn layers (WL, a domestic breed) and red junglefowl (RJF, the ancestor of domestic chickens). Individuals were selected to be homozygous for either the RJF or the WL genotype at a growth QTL previously found to influence several domestication related traits. In the first experiment, we included a comparison between pure red junglefowl (17 animals) and White Leghorns (20 animals). Data were analysed with ANOVA and t-test. WL birds spent more time inspecting an unfamiliar individual at four weeks of age than did RJF birds (55 ± 6 vs 16 ± 6% of time; P<0.001), and male WL-genotype AIL birds (8 animals) inspected more than RJF-genotype males (5 animals) (33 ± 12 vs 0 ± 0%; P=0.02). Furthermore, adult WL-genotype AIL birds (9 animals) were less aggressive to unfamiliar conspecifics (5.7 ± 4 vs 14 ± 5% of time intervals; P=0.009) than were RJF-genotype birds (11 animals). The results suggest that domestication has caused changes in social behaviour, which, in males, partly depend on variations in the genotype on the growth QTL. This QTL harbours the AVPR1a gene, a receptor of the hormone vasotocin. This gene has been found to influence social behaviour in finches, rats, mice and voles. Although many other genes are located in the QTL, AVPR1a seems particularly interesting for future studies.

Animal welfare aspects of gene technology in animal production

K. Hagen, Europäische Akademie Bad Neuenahr-Ahrweiler GmbH, Wilhelmstr. 56, 53474 Bad Neuenahr-Ahrweiler, Germany

In genetic, physiological and biomedical research, transgenic animals are already commonplace, but nevertheless, very few studies have addressed welfare implications of transgenesis. Improvement of transgenesis techniques and functional genomics have opened up vast application possibilities for genetically modified animals in production, too. Potential future applications include increased production rates, altered products, decreased susceptibility to stress, and reduced environmental impact. Applications also include novel products: nutraceuticals, industrial compounds and biopharmaceuticals. The purpose of this study is to provide an overview of the potential implications of genetic engineering on animal welfare in production animals. Although some systematic research does exist, most of what is known about health and welfare risks of transgenesis is based on anecdotal or circumstantial evidence, primarily related to the research phase (establishing a transgenic line). Important aspects include: i) Some of the reproductive methods associated with genetic engineering (notably, cloning) lead to high abortion rates and health problems in some species. ii) Transgenesis can be accompanied by genetic or epigenetic disruption and aberrant gene expression, with a variety of health and welfare implications. iii) Harm can be caused by the bioactivity of foreign proteins entering the body's circulation, or getting expressed at unintended sites. iv) subtle dysfunctions may remain undetected. Animal welfare research should accompany the development of livestock biotechnology. To ensure early detection of welfare-compromising effects of transgenesis, welfare evaluation schemes should be developed that can be integrated as part of the biotechnological transgene evaluation (such schemes exist for transgenic mouse strains, although they are not used much). Long-term evaluation is needed if transgenic lines enter production, to assess effects that may develop with age and in future generations.

Of mice and pigs: tackling the lack of welfare assessment (protocols) in transgenic farm animals

R. Huber[1], D.B. Sørensen[2], P. Sandøe[3], B. Whitelaw[4] and I.A.S. Olsson[1], [1]IBMC-Instituto de Biologia Molecular e Celular, Universidade do Porto, Laboratory Animal Science, Rua do Campo Alegre 823, 4150-180 Porto, Portugal, [2]Faculty of Life Science, University of Copenhagen, Department of Veterinary Pathobiology, Division of Laboratory Animal Science and Welfare, Grønnegaardsvej 15, DK - 1870 Frederiksberg C, Denmark, [3]Faculty of Life Science, University of Copenhagen, Centre for Bioethics and Risk Assessment, Rolighedsvej 25, DK-1958 Frederiksberg C, Denmark, [4]Roslin Institute and Royal (Dick) School of Veterinary Studies, Division of Developmental Biology, Roslin BioCentre, Midlothian, EH25 9PS, Scotland, United Kingdom

The idea of genetically modified (GM) farm animals is met by scepticism by large parts of society, raising among other issues the risk that genetic modifications result in welfare problems. This issue is very real when it comes to GM farm animals used in biomedical research to study diseases. In potential future genetic engineering of animals for agricultural production, animals will not deliberately be created to develop health problems but compromised welfare may be an unintended side-effect of the genetic interventions which will need to be addressed. While there have been considerable efforts in the welfare assessment of farmed animals and phenotype characterisation is common practice for GM mice, there are no phenotyping protocols for transgenic farm animal species and the most used protocols for mice pay little attention to animal health and welfare. An adaptation of existing methodology for farm animals is needed as the research context is critically different from the farming context: focus is on individual animals rather than groups and on health and welfare problems that arise due to genotype rather than environment. It appears reasonable therefore, to combine existing methods and knowledge from both groups of animals mentioned above and to adapt them to transgenic farm animals at an early stage of research in order to comply with the scientist's responsibility towards the animals and - in the course of public discussion - society. In this presentation, we will discuss how this can be achieved, based on an ongoing project aimed at the development of new tools for genetic engineering of farm animals wherein phenotyping and welfare assessment schemes for two species (mice and pigs) are applied. The potential synergies and differences in their approach between (and within) scientists interested in characterising and using new phenotypes of transgenic animals and animal welfare scientists will be discussed.

Optimisation of reproductive effort in mouse females

A. Dusek[1,2], L. Bartos[1] and F. Sedlacek[3], [1]Institute of Animal Science, Department of Ethology, Pratelstvi 815, CZ-104 00 Prague, Czech Republic, [2]Charles University, Department of Zoology, Vinicna 7, CZ-128 44 Prague, Czech Republic, [3]University of South Bohemia, Department of Zoology, Branisovska 31, CZ-370 05 Ceske Budejovice, Czech Republic

In polytocous mammals, mothers optimise reproductive effort by a trade-off between number and size of offspring. Considering reproductive optimisation, there are two important mechanisms influencing maternal reproductive success. First, mothers regulate offspring number to ensure adequate growth of litter-mates. Second, offspring compete against their siblings in favour of their own growth. To check possible reproductive optimisation, we tested effect of maternal mating condition and offspring relative birth weight on offspring pre-weaning mortality. We studied 137 mothers producing 1840 pups of Charles River CD-1 albino mouse. Maternal mating condition (relative change in weight) was experimentally reduced by exposure to food stress when all food was removed for 24-h every second day during one week before mating (stressed: mean=-9.76%, N=65; control: mean=2.11%, N=72; $p<0.0001$). To estimate the probability of offspring mortality, we applied logistic regression using PROC GENMOD (SAS). The final model contained (i) categorial variable "food stress" ($p<0.01$), (ii) interaction between "relative change in maternal weight" and "birth litter size" nested within "food stress" ($p<0.02$), and (iii) "offspring relative birth weight" ($p<0.05$). To our surprise, offspring of control mothers suffered more from mortality (6.87%) than offspring of stressed mothers (4.51%). The probability of offspring mortality increased with decreasing maternal condition and increasing size of the litter from which the offspring came, being highest in control mothers with large litters. In addition, the probability of offspring mortality increased with decreasing relative weight of the offspring within the litter, indicating that larger pups were more competitive successful than smaller ones. To sum up, reproductive optimisation (1) seemed to be activated by intermittent feeding before mating and (2) resulted both from maternal regulation of offspring number and sibling competition within the litter.

The maternal mediation of fear and stress responses in rodents

H. Würbel, Justus-Liebig-University of Giessen, Animal Welfare and Ethology, Frankfurter Str. 104, 35392 Giessen, Germany

Based on recent research in rodents, developmental plasticity of hypothalamus-pituitary-adrenal (HPA) and fear responses in mammals has been proposed to be mediated by environment-dependent variation in maternal care. According to my own group's research on mice and rats, however, variation in maternal care alone fails to explain the observed variation in HPA and fear responses, while adding environmental stress as a second factor to the model predicted the observed phenotypic changes much better. In this two-factor model, maternal care and environmental stress exert independent, yet opposing, effects on HPA reactivity and fearfulness in the offspring. This accounts well for the finding that completely safe and stable, as well as, highly stressful maternal environments resulted in high HPA reactivity and fearfulness compared to moderately challenging maternal environments. Alternatively, however, developmental plasticity of HPA and fear responses might be a function of maternal HPA activation. Thus, implying a U-shaped relationship between maternal glucocorticoid hormones and HPA reactivity and fearfulness in the offspring, increasing levels of maternal glucocorticoid hormones with increasing environmental adversity would explain the observed developmental plasticity equally well. This raises the possibility that variation in maternal care is an epiphenomenon, rather than a causal factor, in developmental plasticity of HPA and fear responses in rodents. I will examine these two hypotheses based on evidence from my own group's research and from the literature, and discuss their implications for the housing and welfare of laboratory rodents and the limitations of generalisations to other species or taxa.

Maternal behaviour and litter loss in laboratory mice: an experimental and epidemiological approach

E.M. Weber[1,2], B. Algers[1] and I.A.S. Olsson[2], [1]Swedish University of Agricultural Sciences, Department of Animal Environment and Health, P.O. Box 234, SE-532 23 Skara, Sweden, [2]Institute for Molecular and Cell Biology - IBMC, Rua Campo Alegre 823, 4150-180 Porto, Portugal

Mice give birth to large litters of altricial young dependent on their mother for nutrition and thermoregulation, making maternal behaviour a key-factor in offspring survival. A common problem in laboratory mouse breeding is high mortality during the first days postpartum. Even though mice have been used as model animals in laboratory studies of infanticide, few studies have addressed factors underlying litter loss under normal animal facility conditions. When examining the survival rate of 539 litters of two commonly used laboratory strains (C57BL/6 and BALB/c) bred under normal husbandry procedures, we found an average mortality of 28.9% (whole litters lost). Comparable mortality was found in two experimental studies (average mortality 20.6%) and in two breeding colonies (average mortality 35.6%) of mutant and wildtype C57BL/6 mice, with one of the experimental studies showing higher survival in furnished compared to barren cages (Fisher's exact test, P=0.033). All together, these data confirm high mortality in laboratory mouse breeding. To minimise litter loss it is recommended to avoid handling the female and the litter shortly after parturition. However, in an experimental study 49 C57BL/6 and BALB/c female mice were handled and separated from their litter once daily from day 1 postpartum without litter loss being induced. It is still not clear whether the high frequency of litter loss is due to reduced maternal abilities and/or increased stress sensitivity of the mother. With the aim of increasing the understanding of the underlying factors, home cage behaviour after parturition is presently being analysed in a case-control study. Preliminary results show that any litter loss takes place before day three postpartum and there is no indication of female mice actively killing the pups. With an epidemiological approach to examine the present situation in mice breeding facilities in Sweden we aim to reveal factors that could be associated with high litter loss.

Influence of genetics and environment on maternal behaviour and piglet survival

E.M. Baxter[1], S. Jarvis[1], R. Roehe[1], A.B. Lawrence[1] and S.A. Edwards[2], [1]Scottish Agricultural College, SLS, West Mains Road, Edinburgh, EH26 0PH, United Kingdom, [2]Newcastle University, School of Agriculture, Food and Rural Development, Agriculture Building, Newcastle upon Tyne, NE1 7RU, United Kingdom

In order to improve piglet survival in sow-welfare-friendly systems, genetic strategies could be adopted. However, an understanding of the characteristics of sows and piglets that influence survival in alternative farrowing systems is essential to develop new breeding indices. Sixty-five gilts and their piglets (757) from two genotypes, High postnatal Survival (HS) or Control (C), were studied in both indoor loose-housed and outdoor farrowing systems. We measured behavioural and physiological indicators of pre-weaning survival in gilts and piglets in both farrowing environments to determine the influence of genetics and environment on potential survival indicators and survival per se. GLMM analysis determined the difference in mortality between the genotypes. Genotype affected the risk of total mortality at a piglet level in the outdoor system (C: 17.9% (±3.23) vs. HS: 12.2% (±3.46), W_1=3.60 P=0.058), but there was no effect in the indoor loose-housed environment (C: 12.3% (±2.69) vs. HS: 14.9% (±3.18) W_1=0.07 P=0.797). GLM analysis determined gilt differences. Genotype influenced maternal characteristics, with C gilts in both environments being significantly more careless when changing posture during farrowing and exhibiting more crushing behaviour (mean deviance ratio$_{1,62}$=47.25 P=0.002) than HS gilts. Since reduction in crushing is a major justification for the confinement of sows in farrowing crates, this is an important trait if promoting the use of alternative systems. However, only HS gilts showed piglet-directed aggression (mean deviance ratio$_{1,62}$=64.90 P<0.001) and this was predominantly in the indoor environment, suggesting a genetic effect on environmental sensitivity. The relatively low mortality of HS piglets in this study indicates the potential to use genetic selection strategies in less restrictive systems to benefit both piglet and sow welfare. However, the genetic by environment interaction in aggression highlights the complexities of such strategies and the importance of behavioural, as well as physiological traits when developing new breeding goals.

Are non-nutritive nursings a good maternal behaviour in domestic pigs?

G. Illmann and M. Spinka, Research Institute of Animal Science, Department of Ethology, Pratelstvi 815, 10401, Czech Republic

The milk transfer from the mother to the young is the most important parental investment. Non-nutritive nursing, where the young is making nipple contact and sucking but without milk being transferred, is a widespread phenomenon among mammalian species. In most cases, non-nutritive nursing occurs as a part of a nursing bout. However, in species like domestic pigs nursings can be completely without milk ejection. In domestic pigs up to 30% of nursing bouts do not result in the let-down of milk. What biological function may the non-nutritive nursings have? In this study, we suggest four answers to this question and review the evidence in their support (i) non-nutritive nursings regulate the milk production. (ii) non-nutritive nursings suppress the mother-piglet conflict over the milk distribution. (iii) non-nutritive nursings suppress allo-suckling (iv) non-nutritive nursings result from disturbance and stress and can be a welfare problem in modern housing systems. We found evidence that sows through non-nutritive nursings control the maternal investment in the current litter (i and ii) and that non-nutritive nursings decrease the non-offspring nursing in the group of lactating sows (iii). There is no clear evidence that non-nutritive nursings are a result of disturbance or stress.

Authors index

Hemmings, A.	152
Hemsworth, L.M.	147
Hemsworth, P.H.	42, 75, 117, 183
Henshall, J.M.	36
Hepola, H.	88
Hernández, A.	202
Hernandez, C.E.	77
Herrera, J.G.	200
Heyman, Y.	85
Higginson, J.	18
Hild, S.	78
Hillmann, E.	106, 174
Hinch, G.N.	45, 206
Hirahara, S.	184
Hockenhull, J.S.	222
Hogan, D.	194
Holcombe, S.J.	29
Horn, T.	16
Hörning, B.	164
Horsberg, T.E.	13
Hothersall, B.	87, 155
Hötzel, M.J.	168
Houbak, B.	79
Hovland, A.L.	46, 188
Huber, R.	246
Huber-Eicher, B.	57
Hurley, J.	218
Huzzey, J.	163

I

Iacobucci, P.	130
Illmann, G.	127, 251
Irwin, R.M.	230
Ishiwata, T.	27
Ison, S.H.	33
Ito, S.	134, 179

J

Jack, M.C.	33
Janczak, A.M.	13, 194
Jarvis, S.	48, 250
Jauhiainen, L.	88
Jekkel, G.	201
Jensen, M.B.	40, 175
Jensen, P.	21, 35, 80, 92, 244
Jeon, B.S.	121, 123
Jeon, J.H.	121, 123

Johnson, A.K.	191
Jones, B.	119
Jones, B.R.	214
Jones, M.A.	28
Jones, S.	38
Jongman, E.C.	143
Jöngren, M.	35
Jørgensen, E.	79
Jørgensen, G.H.M.	34, 133
Juste, M.C.	215

K

Kaart, T.	97
Kaleta, E.F.	59
Kanerva, K.	221
Kanitz, W.	172
Karlen, G.A.	75
Karlsson, A.C.	21
Kasuya, E.	134
Keeling, L.J.	3, 112, 154, 216, 220
Keil, N.M.	65, 90
Kenny, D.A.	93
Keppler, C.	24, 59
Kerje, S.	21
Kielland, C.	226
Kiley-Worthington, M.	6
Kilgour, R.J.	27
Kirkden, R.D.	188, 191
Kissinger, C.	126
Kistler, C.	66
Kittilsen, S.	91
Kjaer, J.B.	185
Kjellberg, L.	150
Kluivers-Poodt, M.	17
Knierim, U.	24, 59, 187
Knowles, T.G.	23
Koene, P.	20, 47
Kohari, D.	100, 122
Komen, J.	30
Kondo, S.	103
König, B.	66
Kosako, T.	100
Kotrschal, K.	69, 146
Krebs, N.	62
Krohn, T.C.	228
Krösmann, K.	58
Kruschwitz, A.	57

256 Applied ethology

Printed in the United States
by Baker & Taylor Publisher Services